THE MAGIC
OF CODE

ALSO BY SAMUEL ARBESMAN

The Half-Life of Facts: Why Everything
We Know Has an Expiration Date

Overcomplicated: Technology at the Limits of Comprehension

THE MAGIC
OF CODE

How Digital Language
Created *and* Connects Our World—
and Shapes Our Future

SAMUEL ARBESMAN

PUBLICAFFAIRS
New York

PublicAffairs
Hachette Book Group
1290 Avenue of the Americas, New York, NY 10104
www.publicaffairsbooks.com
@Public_Affairs

Printed in the United States of America

First Edition: June 2025

Published by PublicAffairs, an imprint of Hachette Book Group, Inc. The PublicAffairs name and logo is a registered trademark of the Hachette Book Group.

The Hachette Speakers Bureau provides a wide range of authors for speaking events. To find out more, go to www.hachettespeakersbureau.com or email HachetteSpeakers@hbgusa.com.

PublicAffairs books may be purchased in bulk for business, educational, or promotional use. For more information, please contact your local bookseller or the Hachette Book Group Special Markets Department at special.markets@hbgusa.com.

The publisher is not responsible for websites (or their content) that are not owned by the publisher.

Library of Congress Cataloging-in-Publication Data has been applied for.

ISBNs: 9781541704480 (hardcover), 9781541704503 (ebook)

LSC-C

Printing 1, 2025

To my parents, who first introduced
me to the magical world of computing

CONTENTS

Introduction

Computing as the
Supreme Connection Machine

C omputing is a wonder of the world, standing alongside the Great Pyramid and the Hoover Dam as something worth marveling at, our jaws agape and minds expanded. Chances are, though, you don't feel this way. Computers and their powers have receded to a sort of background noise, an ever-present technological scenery. Or we look, many times rightly so, at computers with worry and concern, anxious of their powers and their hold on us. Wonder does poke through from time to time, when we notice that, for instance, artificial intelligences have started to write poetry or can generate images by request. We are then forced to acknowledge that, yes, at least at some level, computers are impressive.

But for me, this impressiveness is not the only—or even the most important—cause for wonder when it comes to computing. In the same way that science fiction is not just about whiz-bang gadgetry or swooping starships but also about engaging with mind-boggling ideas and thought experiments, the wonder of computing is not just the cool technologies we see around us. The wonder is that learning about computers—their nature, their history, and the snaking tendrils of their impact—leads to marveling at basically everything around us. Understanding computers and code and computation—in all their weirdness and delights, features and implications—can lead you to think about so many aspects of our world, from biology and life itself to how we use language and how we think. It is the ultimate connector.

The novelist Richard Powers has described the novel as a "supreme connection machine." Computing can be the same thing: when you take it seriously, it's a massive connective engine to everything around us and what we know, acting like a centripetal force that draws together so much of our knowledge and is cause for a kind of panoptic wonder. While we might relegate computer science to a corner of engineering, it is much bigger and unifying, swirling around, drawing in, and touching every other way of thinking and area of knowledge.

———

In his book *Philology: The Forgotten Origins of the Modern Humanities*, James Turner claims that, before there were fields of history, linguistics, anthropology, and other domains of the humanities, there was philology, a sort of ur-field, similar to how natural philosophy preceded all of the scientific disciplines that we know and love. Philology was the "king of the sciences," examining the origins and etymologies of words and languages but in the process ranging over linguistics, history, archaeology, literature, theology, art, and more.

Philology was a sort of intellectual attractor, using the study of language to draw a huge number of topics into its orbit. J. R. R. Tolkien was a philologist, and, while no doubt an outlier, *The Lord of the Rings* is the kind of wide-ranging and all-encompassing work you can create if you take your philology seriously. But as the study of different humanistic fields expanded and deepened, there was a fracturing and specialization. As historians, literature professors, and archaeologists each carved out their niche, philology slowly faded away from the humanities.

The study of computing is in many ways similar to philology: computation is the same kind of universal solvent, absorbing and melding with many other domains. Computation is not simply a subdiscipline of mathematics or science but almost a liberal art that can range across and interact with other fields.

This is due to the particular nature of code and software. Computer code is not concrete, it's not steel, but it's also not just text. Software consists of spells of crystallized thought. When a programmer writes code, she constructs a model in her mind of how a system works, or how something in the world works, and embodies that mechanic into computer logic. When my daughter works on rudimentary coding projects at school in her technology classes—programming a digital monkey to collect bananas on the screen, for instance—she has to imagine herself as the machine, determining how it operates and how a loop might work. This process of precipitating whatever was imagined by the programmer into an object of language and action is what brings us into an almost magical realm: for the computer, logic is not inert; it is set to running and is given a life of its own.

In fact, once you scratch on the idea of software as the stuff of language marinated with power—scraping your thumbnail over the gray substance of that lottery ticket of technology—so much is revealed. If you study computation, you are not simply thinking about the nature of loops, data structures, or databases. By thinking of computation broadly and grappling with its profound ideas, you are able to interrogate so much about the world around us. Software is capable of revealing connections to biology, human language, physics, society, and even philosophy. Massive simulations can be built in silico that span civilizations, contain the interactions of a million stars, or even embody the properties of life itself. The complexities of human communication are revealed in large AI systems that chat with the user or translate from one language to another. When we take this wide view toward the nature of code, we can see that its deep history touches on everything from machines to myth. In fact, code is the embodiment of the fantastical stories we have been telling ourselves for millennia that speak of controlling the world through our words. Code is spellwork and magic made concrete.

When we think of computation, then, we must recognize that it touches on all aspects of the human condition. However, most of computing's implications have occurred in this post-philological fracturing: we employ code for specific uses and in different subspecialties but rarely think of it as part of a unifying, all-encompassing domain of thought. But computation is beyond computer science, or computational biology, or any other field developed by tacking "computational" onto some other word. This broad sense of computation is a smashing together of computer science, the humanities, and the sciences both natural and social.

Think of this book as a guide to broadening our understanding of the wonders and weirdness of computation, as a vehicle for understanding the history and nature of computing and its relationship to us as humans. For only when we understand the true breadth of computing can we figure out how best to incorporate it into our lives, making it work for us, rather than the other way around.

This will necessarily be a personal and subjective exploration of some of my favorite scenic views, as it were, in the wide world of computing, and is organized into three sections—Code, Thought, and Reality—with each one spiraling outward in impact and implications.

———

What is in store for the reader? After first exploring the nature of wonder itself in computing, Part I, Code, begins by examining what code is—this strange stuff that we use to control our computers—and how it works. After this exploration into code, we move on to what it means to take the magical nature of computation seriously: If coding really is like sorcery, what does this mean for how we think about computers? Is this a fruitful analogy that can teach us something about both code and our relationship to it?

With this foundation laid for understanding code itself, we move to two different aspects of open-ended software creation. The first focuses on how software is made, specifically open source software. It turns out that the world of open source is a kind of textual tradition, and if we view it this way, we can better understand how computer programs are built. The second aspect of open-ended computing is an exploration of one of the quintessential properties of computer programs: that if you do lots of little calculations and operations over and over, you can get something qualitatively new and beautiful. Specifically, by taking this feature of computers seriously, you can write programs that combine mathematics and computing, and from this seeming simplicity unfurl complex and surprisingly rich microcosms. Entire digital universes emerge from computation, from infinitely deep fractals to tiny computer programs that generate artworks, each one different every time they are run.

Part II, Thought, begins to examine how we actually use code and how it impinges on our thinking. The first chapter of this section is all about language, both natural and computational. Humans are a linguistic species, but so are machines: we interact with them using a variety of languages, and so understanding the nature of programming languages and the ways that they are both similar to and profoundly different from natural language is vital. I then explore what it means when we are able to let everyone, not just a computational elite, code—when each of us can imagine computer programs, take these ideas in our minds, and encapsulate them in software. This democratization of programming, from spreadsheets to AI tools, changes how each of us can think about our relationship with computers. And finally, in this section's last chapter, I directly confront thought itself: What does it mean to build computational tools for improving our thinking? I look at the interaction between thought—this most human of tasks—and machines and also focus on the new artificial intelligence software that is allowing these interactions to blossom.

Part III, Reality, completes the spiral outward into how code impacts our view of the world and the entirety of the cosmos. I first explore the powers and perils of computational simulation, the process of modeling our world inside computers. The next two chapters examine different aspects of biology: the ways in which computing and biology are similar, different, and increasingly colliding, and then what it might mean to create artificial life and encapsulate features of biology within computers. We then examine glitches and errors in computers, and how this is intertwined with the fact that computers are deeply physical objects; they are not disembodied spirits but machines made of metal and sand. After this exploration of the physical nature of computing, we finish this section with the simulation hypothesis: What if all of reality is but a computer program, and each of us are part of this massive piece of software? What would this even mean, and what might it mean for our humanity? And can you still learn something interesting about computers and the world, even if you don't take this idea that seriously?

I conclude with a bit of philosophizing, a summation but also a sort of reckoning with how to be human in this age of computation. You will find that throughout the book there is a drumbeat of humility—what we can know and what we cannot, what we can do with machines and what might be forever beyond our grasp—and it also appears in this concluding chapter.

While this book is aimed toward those with no programming experience (yet!), I hope seasoned developers might also get something out of it, both due to its broad scope but also because of what I hope is a somewhat distinctive and novel perspective on the world of computation. Whether or not you've experienced computer code firsthand, my sincere goal is that this book becomes a compass-like guide to this broad humanistic perspective, that there is grandeur in this view of computation.

It is a love letter to the computer, in all its glory and implications.

———

The biblical scholar and translator Robert Alter has noted that ancient Israel was far from a significant player in its world, surrounded by the major empires of Mesopotamia and Egypt and buffeted by their geopolitical machinations. These empires had a physical splendor—ziggurats, pyramids, temples—that the world of ancient Israel could not hope to match. And yet this tiny country produced some of the most timeless and beautiful literature in all of history.

I wonder if there is a larger point here: an enduring edifice can be built in many ways. Some build pyramids or skyscrapers. Others construct worlds from words. The primacy of text and its subtle sophistication—whether to convey ideas, tell stories, teach lessons, or, now with code, act in the world and process information—is something that we need to rededicate ourselves to. Textual splendor is no less an accomplishment than physical richness. But we need to pay attention to the power of text.

In the book of Genesis, Jacob—soon after fleeing his home to avoid his brother's murderous rage—falls asleep on the way to his mother's family. Upon waking from a dream of prophetic wonder, Jacob declares the awesomeness and majesty of his location, "and I did not know it!"

We are surrounded by the power of code and computation, weaving a future of unparalleled emergent strangeness and wonder. It's time we know it.

———

Let's begin with wonder.

1

Computational Wonder

Entering the Garden of Computing Delights

The computer—at least its parts and technologies—is on a continuum with many advances that came before it. Mechanical devices had been built for arithmetic hundreds of years before the twentieth century. The nineteenth century and early twentieth century saw a burst of clockwork-like mechanisms designed for such calculations. And components of the digital computer itself—from vacuum tubes to wires—already existed, though were used for other purposes. Nevertheless, when all of this came together, something qualitatively new was created: the modern computer was distinct from what had come before it. But many features of this difference have only slowly been recognized, taking nearly a century for us, as a society, to unspool the implications of this idea of computing.

Intriguingly, though, the features of computing that connect it to so much and make it feel like a liberal art have been with us since its origins. If we look at the development of Stanford University's computer science department in the mid-twentieth century, it emerged from the mathematics department but also touched on early work in artificial intelligence and chemistry. The computer science department at the University of Michigan was said to have grown out of its philosophy department. Early in the history of the digital computer, experiments on digital "organisms" within a machine were already being run, and almost as soon as the computer was invented it was being used for large-scale simulations that spanned the entire planet.

However, while the seeds of the all-encompassing nature of the computer were evident almost from the start—if you looked for them—it has taken decades for these seeds to mature. To borrow another plant metaphor, the tendrils of computation extended far and wide, but they needed time to thicken and grow in order to touch so many different domains.

The history of computing—as we will see—is one of unearthing these veins of computational wonder and tracing how they have moved from the subterranean depths of our experience and slowly bubbled upward, infiltrating more and more of our lives and seeping into the world. Alongside all of this are two mindsets—utilitarian and wondering—that have coexisted throughout the history of computing, sometimes intersecting but more often eyeing each other warily across a chasm.

To understand them, we must examine the idea of enchantment, the ways we have thought about wonder throughout history.

COMPUTING ENCHANTMENT

In Philadelphia's Franklin Institute science museum there is a small, brass, doll-like torso attached to a wooden box and table: a mechanical automaton. This machine is designed to draw various images. You put a pen in its right hand and, following its gear-like memory disks—their rough edges contain its "memory"—it will create one of a handful of images, from a sailing ship to short poems. This was all made over two hundred years ago and is an engineering work of art and wonder, using clock-style mechanisms to produce behaviors that appear to require intelligence.

But during this same era, there were other engineered automata that were a bit less "real." In the eighteenth century, a mechanical duck automaton was developed, a self-operating device with an internal mechanism that allowed it to appear that it was digesting food it consumed. It didn't actually do this: there were separate compartments for food intake and for excrement. But many observers were impressed by the appearance of this feat. Arguably the most famous of all mechanical automata was the Mechanical Turk, which could play chess at a high level. But it, too, was a fraud, with a person hidden inside the machine to operate it.

For hundreds of years, there was a tradition in Europe of devices known as "brazen heads": bronze magical or mechanical heads that could answer questions and chat with the user (one from the thirteenth century was apparently so talkative and annoying that it was destroyed). There is no evidence that these brazen heads were real.

These stories imply something intriguing about this early modern era: despite society being in the throes of the Enlightenment, technology was still mingled with magic, and humbuggery with possibility. The society of that time and place was still working out the boundaries between reason and irrationality, what was truly possible with technology and what was not.

However, we have since passed this time and are firmly in the realm of disenchantment, a shift in Western civilization described by the German sociologist Max Weber. Beginning with the Enlightenment and accelerating through the Industrial Revolution, the argument goes, the process of disenchantment removed any place for magic—or even supernatural religion—within society, replacing it with science, reason, logic, and a resolutely secular world. The days of demons and elves, witches and wizards, were behind us.

Of course, this is not quite true. We are still surrounded by enchantment. We might believe in ghosts and angels, or in the power of kabbalistic red threads worn by Madonna. Modern humanity is still steeped in irrational thinking, from the paranormal to QAnon. In fact, as Kurt Andersen lays out in his book *Fantasyland*, not only is this enchantment still with us, but it is a constant thread in the history of the United States, from before its founding with the Puritans and their witch trials to a cornucopia of New Age ideas more recently.

More properly then, we are in a world where, according to the philosopher Charles Taylor—whose magisterial work *A Secular Age* explores these changes over the past several centuries—we went from enchantment as the default mode five hundred years ago to it being simply one among many options in our modern world. We now have a buffet of worldviews from which to choose.

When it comes to technology and science, however, we are supposed to be immune to such claptrap. For those of us who reject all of this, the world does indeed seem disenchanted. There are no gods to call upon or witches to provide us with amulets. We can't indulge in magical thinking.

When we look at the history of computation, we see something of this disenchantment, though of a different sort. I do not mean a belief in irrationality or something supernatural. Rather, there is a sense that there has been a loss of wonder when it comes to certain aspects of computation and software. This "wondering stance" has vanished from our approach to computing. We have replaced a sense of the sublime and delight with one of utility and hard work.

Previous decades of computing seem filled with analogies to magic and programming wizardry, as we will see, but it has given way to the logical and streamlined process of corporate software engineering. Programming by bureaucracy is a far cry from referencing the "Wizard Book," the nickname of *Structure and Interpretation of Computer Programs*, a foundational computer science textbook used widely to provide an introduction to computing (including in my own college experience).

How did this happen? There's a kind of folk history for this that can be told that primarily comes down to scale, explaining how we moved from playful hackers to the cubicle-based software engineers required for the creation of software by large teams. As computers and operating systems became more sophisticated, they required more coordinated effort, turning projects into works of engineering, not feats of wizardry. The work of software creation went corporate, denuded of any such playfulness: in order to enable a large group to work toward making a slick product, software teams had to abandon this feeling of sorcery. Developing an application for a personal computer does not feel like magic; it feels like work, with project milestones and Slack channels and brief daily stand-ups. This can be good for producing these commercial applications, but in the process something has likely been lost.

Sorcery, on the other hand, should be more solitary, haphazard, and playful. We see this in the hackers of decades ago, who proudly wrote arcane and powerful computer code; Steve Wozniak wrote the implementation for Apple's BASIC language on paper entirely by hand. These were swaggering cowboys of the command line. Overall, the inability to work individually on a piece of software often feels like it has changed our experience of this world, even just for its users.

This is part of a disenchantment with technology more broadly: we have become captives of Big Tech, the corporations who do everything in their power to build mass-market software devoid of any whimsy, from Google Docs to Microsoft Teams. We have coding academies, draining the delight from software and turning it into a method of mass production. We have forgotten that computing should be viewed with wonder and a sense of digital alchemy. Or if we do indulge in magic in the realm of technology, it is more along the lines of magical software that "just works." Or the desire to utter a benediction to the gods of the cloud when searching for a document that, pretty please, has been successfully backed up and not lost forever.

The history of computing is, however, more complicated, messy, and interesting than my simple narrative. The earliest computers—those room-filling monstrosities—were found in universities and corporate environments and were necessarily complex and team-oriented affairs. Programming these machines in the 1950s and 1960s was not just the province of the playful individual but involved acting as a component of a larger behemoth. There were hints of a different, more enchanted approach, such as the hackers during this time at MIT, including the creators of the early video game *Spacewar!*, which was developed for the Digital Equipment Corporation (DEC) PDP-1 "minicomputer"—smaller than a mainframe but still far larger than a personal computer, a sort of evolutionary pit stop on the way to personal computing.

Once personal computers were introduced in the second half of the 1970s, a greater space for this wondering stance was created by exposing everyday users to the world of computing. But alongside the personal computer revolution was the professionalization of software development—the utilitarian stance. Building software was the endeavor of companies like Microsoft or Lotus, a spreadsheet and business-tool maker. By the late 1980s, software development could be managed like a construction project, with massive charts hung on walls and progress tracked in detail, with very little room for quickly adapting to changes or problems that arose over the course of the project (I'd like to think by now we've realized that making software should be different from a manufacturing process that cranks out widgets).

But alongside the manufacturing or construction model, there were other ways of making software. My family has been using Apple Macintoshes since the late 1980s, and in addition to the software we had that was made by Microsoft, Apple, or other large companies, there was an entire world of software coming from small, independent developers and teams. These programs were wackier and stranger, ranging from little games to tweaks of your operating system that would, for example, make Oscar the Grouch emerge out of your digital trash can and sing his love of trash. I loved these kinds of programs when I was younger.

In the modern era, we have individual creators and artists building software that coexists with the tech titan conglomerates—just as we don't have to choose between industrial agriculture and having a backyard vegetable garden or between frequenting chain restaurants and cooking at home. The novelist and writer Robin Sloan has written of the delight of creating your own artisanal software as a kind of "home-cooked meal." For example, Sloan built a smartphone app designed to share photos and videos, but just for his own family members. It was a home-cooked piece of software for his family. Large corporate efforts could sometimes yield a sense of wonder too: for example, the McDonald's website from 1996 is adorable, with graphics that appear hand drawn and the web design sophistication of a grade-schooler.

Enchantment and the belief in nonsense never disappeared in our modern scientific age. Similarly, the worlds of enchantment—or really, wonder—and disenchantment have existed alongside each other in the realm of computing. The utilitarian road of staid and corporate coding has always run beside the path walked by those delighted by the marvels of these machines, those who have played with computers at a more human scale. In fact, a lot of the human-scale computational wonder might only be possible because it is built on top of hardware and software created using a utilitarian stance. There is a bit of a symbiosis here between these two approaches.

Many of us have had these experiences of wonder with computers, even if we have forgotten them. For me, one such memory was sitting as a kid with my father while he entered BASIC computer programs into our Commodore VIC-20 personal computer. He'd type a program from a magazine, hit a key, and a pixelated game would unfurl across the screen. A few years later we got an early Macintosh computer, and that giddy feeling compounded. There was the smile-like slit on the front face of the computer: its disk drive, able to swallow your disk whole and spit it out on command. It could talk, with rudimentary text-to-speech, which I had never before experienced. It even had a mouse, which was something rare and special at that time (our Mac came with gamelike software specifically to teach you how to use the mouse properly).

Or there was the experience when my father and I logged onto the pre-Web Internet using an early modem, heard a series of beeps, whistles, and crash of static, and then downloaded a text from Project Gutenberg. Or the first time I built my own website on

GeoCities, learning the angle-bracketed runes necessary to generate everything from a table of rows and columns to blinking lines of text. These and other moments like them validated the belief that software and computation could be a source of delight.

It is a conscious choice to choose the path of wonder. But I think we can make an even stronger claim. Not only must we seek out and choose this wondering stance—which can teach us about many aspects of the world—but in doing so, we can learn more about how to be the best versions of ourselves, to even reclaim some of our humanity that many feel has been lost in the march of technological progress.

You might feel that we outsource too much of our humanity to machines, or simply that we become horrible versions of ourselves when we spend time online. But these are the paths of least resistance: we do what is easiest for us to do, because the companies that have designed the hardware and software have incentivized these kinds of behaviors. Or if we really want to be glib, it's utilitarianism, but in terms of the utility that we, as people, are providing to the world of computing. We humans are the ones being used by these technologies and their creators.

In the process of appreciating the wonder of computing, not only can we shed light on so much else around us, but we can be more deliberate about how we relate to computers and what we choose to have humanity mean. This is the true end of the spiraling aspect of computation as humanistic endeavor: it touches on so many different aspects of our world, but this should be in service of helping us to become better people, whether in understanding ourselves or in acting toward each other. Admittedly, this is a very large task—advocating wonder toward computation, showing how it impacts numerous different domains, and providing a lens for thinking about us as people—but if computing is omnipresent, it's at least worth trying.

THE CURRENT
COMPUTATIONAL MOMENT

We are in a particularly critical moment in time when it comes to software and its impact on the world. Modern digital computers are less than a century old, and their powers have been growing exponentially the entire time. Certain philosophical thinkers who focus on the risks to human existence even speak about this time as a kind of "hinge of history," with our current moment's importance due at least in part to the rapidly accelerating powers of our machines. If we are in this special time of wonder, not only are we lucky to be alive right now, but it's imperative that we be aware of software's novel way of intersecting with our world and what might yet be possible.

But we've forgone this opportunity. Surrounded by ever-more powerful technologies, we have been conditioned to view them not as the product of software wizardry but as no more than baubles meant to capture our attention and our dollars. Even though these wonders still exist around us, we too often fail to recognize them as such or don't even notice them at all.

One seemingly dazzling advance right now is that of artificial intelligence, an issue I will focus on throughout this book. Can I predict what will happen in the coming years? No. I am constantly surprised by what is possible, as well as how quickly we have adjusted to the changes we have seen already. We effectively blew past the Turing Test—as commonly understood, an assessment based on a 1950 article by the mathematician Alan Turing that revolves around being unable to distinguish between a computer and a human after a short yet wide-ranging text chat—in late 2022 with Open-AI's large language model ChatGPT. But instead of marveling at this fact, we spent much of our time arguing about how worried we should be about malfunctioning chatbots. If you want to worry about job replacement and the future of work, or how it might affect critical thinking—and I do—that's reasonable, but that worry should at least be partly mingled with wonder. Decades of computer science research have resulted in these profound achievements, and it's worth celebrating, alongside all of the valid concerns.

Of course, this is the way of the world and what happens with technology more broadly. We yearn for an advance, see an early and incomplete version, and quickly move from impressed to disappointed. Then, when a better version finally arrives, sometimes we adapt to it so quickly that we forget we might have ever been surprised by it. It's the hype cycle mingled with a hipster affect: we forget the first flowering of excitement—followed by rapid disappointment—and blend it with an attitude of forever remaining unimpressed.

But send your thoughts into the mists of your technological past. I remember when it was possible to imperfectly remember a commercial from your childhood and not be able to summon it instantly. Or to argue over a half-recalled fact and not be able to determine its veracity right away. Or to arrange to meet friends somewhere and have to be there, because no last-minute messages could be sent. Equanimity in the face of never seeing that advertisement again, or not knowing some fact, or being unable to contact your friends was expected. Then YouTube, Wikipedia, and smartphones swept these experiences away. And they are marvelous inventions (usually). So let's appreciate them for a beat.

We must allow space for wonder and curiosity when thinking about AI, in addition to all the valid concerns. These are not either-or. Being worried about a technology can be rational. My wife and I have had many conversations, both together and with our fellow parents, about smartphones and their effects on children, for example. All of these technologies have a time and a place, and using them everywhere and for everything is an abdication of human choice and responsibility. But not being able to recognize certain aspects of computing's radical difference from what has come before it? Not being surprised and delighted by it? Come on. Yes, an undifferentiated, pessimistic stance toward the new and the different allows you to hang out with the cool kids, but constantly uttering, "Over it," just because AI hallucinates or stumbles on logic problems? Grab yourself and shake hard. That unexcited stupor doesn't do you any favors.

The counterpoint here is that all of these computational advances—which seem so grand and earth-shattering—might be relatively small so far. Humans are great at adapting to change when the changes are small or the stakes are low. But when there was last a vast, destabilizing change—the Industrial Revolution—it did take us much longer to adjust than we would care to admit. As the Internet scholar Clay Shirky has argued, the changes wrought by the Industrial Revolution, particularly urbanization, were so shocking that an entire generation of Londoners self-medicated with an ocean of gin before they figured out how to handle them. As technological advances accelerate, particularly in AI, we must do better than creating a Gen Gin.

And what of code? Some might say that with all of these AI technologies we don't even need to learn about coding any longer, something that will certainly disrupt the world of software. I don't think that will be entirely true for a long time to come, if ever. But separate from that debate, understanding the relationship between computation and the capacious wonders of the world around us—not simply knowing how to code—is going to be more vital than ever. Computing is a window into everything from mythology and history to philosophy. Furthermore, as we'll see, AI is only one in an ever-lengthening series of tools we have built to ease the distance between humans and machines, something that has been happening for decades. It is part of a much larger story.

The rest of this book will be a guide to these many changes, as well as to what is unchanging in how to think about the world of technology. But let's first begin with examining code itself.

PART I

Code

2

Algebra and Fire

What Is Code?

For a long period of time, hundreds of years ago in the Western world, wilderness and unspoiled nature were not objects of veneration or wonder. They were blights upon the scenery. Mountains despoiled our vistas, and the untamed outdoors was a pox upon the earth.

Attempt to range your mind back toward this moment, where mountains were not the pinnacle—pardon the word choice—of the landscape but objects that ruined our views, described as "warts" or "boils." Humanity's goal was to avoid the wilderness, and our instinct was to recoil before it. To behold a mountain was not to be humbled, or to feel your heart leap, but to be mortified and disgusted.

This is almost impossible to imagine and entirely foreign to our modern view, where we enjoy nature, spend time in it, paint it, look at it, use our vacations to escape into it. But prior to the seventeenth and eighteenth centuries, such was the case in Western society. There was a mortal fear of the natural state of the world. So greatly did this part of humanity shrink from untamed nature that some travelers even asked to be blindfolded as they crossed the Alps, lest they get a glimpse of terrifying wilderness.

But beginning several centuries ago, writers began to write of the sublime, scientists examined mountainous regions, artists began to depict these outdoor scenes, and slowly our views changed.

When it comes to computation, we are stuck in the mindset of hundreds of years ago. When seeing a snippet of code, many nonprogrammers view it the same way, looking away or allowing their eyes to glaze over. They can't focus on it, or at least opt not to. Computer code is an object of fear.

I'm not sure all of us will ever revere pages of computer code quite like how we enjoy sunsets or mountainscapes. But this is a goal worth aiming for: we must strive toward appreciating code and the world of computing. If we focus on code and take it seriously, understanding its properties and quirks, it's possible to admire it—even if you're not a programmer—and see some of its beauty.

This will not be a coding tutorial; those are easily found on the Internet. Rather, this is an exploration of the details of code that highlight the quirkiness and strangeness of computing. It is a guide to the computational roadside attractions available as we think about programming our machines.

THE ORIGINS OF CODE

Software differs from any other material that we use to build things. It is at the intersection of the physical and the textual, a technological membrane between reality and words. And software is built from code.

But what is computer code? Code, programming languages, even how computers work: these are all complex matters, and I am in danger here of making sweeping generalizations that miss caveats and exceptions. Nevertheless, simply put, code consists of a series of instructions designed to be run by the machine. There might be loops, branches, objects of intangible abstraction, and a particular grammar to computer code. The logic of the instructions—under what conditions certain lines are run, how a program describes these loops in the code, and more—dictates how the machine will run the program. But only once it is evaluated by the computer does the code actually do anything. Without this step, code is just inert word stuff typed into a text editor or into some sort of prompt. It is simply carefully ordered symbols.

So converting this logical text into action requires a process of interpretation and execution to make it come alive. The Argentine writer Jorge Luis Borges, whose essayistic short stories betray an obsession with the infinite and the fantastical, was prone to highlighting the intersection of algebra and fire: where the logical and rigid meets the fiery and living. In my view, computation itself—this animated logic—is algebra mingled with fire.

And how does a machine go from text to action? Beneath a computer program lies a vast edifice. On good days, it might conjure up the Newtonian phrase of standing on the shoulders of giants, and on bad ones, a Lovecraftian description of paddling about on a dark sea, ignorant of what lies beneath. While most programmers are either ignorant of or don't worry about this fiery, algebraic admixture, a brief examination of the process of converting text to action is worth the effort. Dig beneath the surface and so much is revealed.

Code can only be run by virtue of a descent into simpler kinds of instructions, ones closer to how the parts of a computer chip actually work. For example, a computer program might start as a text file full of lines of code but then is transmuted by a separate computer program into another set of instructions that your machine interprets directly. These instructions are written in the machine code of ones and zeros and specify the most basic of operations that a computer's processor can actually do: adding two numbers together, or comparing two numbers, or storing a number in a digital bin to be used later. Like the individual motions of a weaver, these actions are simple and straightforward. But when you combine enough of these motions together, a spectacular tapestry might be revealed.

There are two kinds of functions involved in this task, compilers and interpreters. The job of a compiler is one of translation: take a set of instructions in one language and convert it into another one. While the input and output languages can be many things, one common type of compiler takes instructions that are written in the structured form of a programming language, such as C, and converts them into a long list of instructions that can eventually be carried out by the microchip, like the example above. This process involves numerous steps, including "parsing" the text—figuring out how the specific words of a program are related and what they are supposed to be doing—followed by translating these words into machine instructions. This process of compilation allows a programmer to write out the logic of her tapestry without having to enumerate a laborious and mind-numbing series of machine code instructions.

The task of the interpreter, on the other hand, is to actually make the computer do something. The interpreter runs whatever it has been given. It might interpret Python code and run the result directly on your computer, or the processor itself might interpret some compiled machine code it has been handed. Eventually, though, the program is run by the processor.

How were these first compilers written? The first widespread compiler—for FORTRAN—was written in assembly code, another language that is essentially a somewhat more human-readable version of machine code. It couldn't be written in a modern high-level programming language with English-like keywords and a specific grammar, because none could yet be used. So it had to be constructed from assembly: these machine instructions that could determine this translation process from a programming language to the operations of a chip.

And how does the chip carry out these instructions? The microchip itself is made up of a mind-boggling number of microscopic switches. These are the transistors whose numbers double every couple years or so according to the wondrous logic of Moore's law. When combined, they allow for these instructions to be carried out.

But whatever the language, we are still left with a strange truth about code: programmers are writing a long list of carefully constructed sentences whose commands are carried out by a sparking bit of sand found within the bowels of a machine. This fact is something that modern-day programmers can only glance at briefly before being blinded. Older coders might have been well aware of these facts, and some actually spent their time thinking about the specific instruction sets of their microprocessor or using their knowledge of binary or assembly language to wring out results from the small amounts of memory and slow processor speeds they had available at the time. Nowadays, many software engineers are far from these roots of the machine. In many ways, this is actually a good thing, because software writing has become so complex and sophisticated that concerning yourself with the details that lurk below the digital surface is the computational equivalent of spending your whole day thinking about how you're breathing: you would simply be unable to think about anything else (and possibly wouldn't even breathe all that well). Modern software is more than complex enough without having to also contend with those details that once consumed older hackers. This can allow a programmer to develop a web application without having to worry about the allocation of memory resources, or to make a program that draws images to the screen without ever having to think about the specific properties of a machine's graphics card.

For this, we can thank the process of abstraction. Computer programming is ultimately the act of taking small bits of structure and combining them together, and then combining them further, until these larger pieces can be used for a specific purpose and the underlying bits entirely ignored. Just as using a bathroom faucet doesn't require me to know the details of where the water comes from or how it has found its way to my home, software is built upon these kinds of interfaces that allow me to not worry about the underlying implementation. I can read and write files using Python without concerning myself with the intricacies of how the bits and bytes of a file might be stored in memory. This is the process of abstraction—new structures are created, and then the details are abstracted away—and it is a fundamental feature of computing. Even the nature of how we store information is a rich and multilevel one. There are straightforward variables that can act as repositories for numbers or strings of text. But there is also an entire bestiary of data structures—specific forms designed to handle more complex and diverse types of information—from linked lists and hash tables to stacks and binary trees, ones that can be bootstrapped from yet simpler kinds of variables.

Just as abstraction is a powerful feature of computing, it is also fundamental to scientific progress. A physicist is able to understand mechanics and motion without recourse to subatomic particles, having abstracted away these details. An economist can contend with supply and demand without thinking about chemistry. The idea of abstraction is far more than just for code and programming; it can even be applied to our ability to understand the world around us.

Some have described computer programming as like playing with LEGO bricks. And it is, in a way. But it's really a lot stranger. It's more like the situation where some of your LEGO pieces might look like normal brightly colored bricks that can snap with all the others, but they might also tell you the number of seconds since the first day of 1970. There can be a huge amount of complexity hidden within each LEGO piece.

The power of abstraction excuses those who work with computers from knowing the physical details of computation. Moreover, the physical details are fungible. While we think about a computer as composed of silicon microchips laced with transistors, this is not the only way to make one. As long as a particular substrate can allow for logic and storing data, you can make a computer out of pretty much anything.

Everything in our machines, from the ability to multiply two numbers to playing with a spreadsheet, is built from the rules of logic first developed by the English mathematician George Boole nearly two hundred years ago. Boole created his approach to logic far before anyone was thinking about electronic computers, but his ideas are deeply relevant. These rules dictate how we might process simple inputs and are far different from our usual mathematical operations of addition and subtraction; they are based on logic. For example, the Boolean logical operator AND is only true (it yields a binary 1) if both inputs are true. So True AND False yields False, but True AND True yields True (1). If you want to flip a bit from False to True, or True to False, on the other hand, you need a new operation, something known as a NOT operator: NOT 0 yields 1, for example.

Another basic kind of switching logic used in making computational decisions is something known as a NAND gate. A NAND gate is a combination of NOT and AND—think the opposite of AND—so its output is only false if both inputs are true; otherwise it outputs true. It turns out that if you can encode this kind of NAND logic into a physical device, then you can build a computer. You don't have to use actual zeros and ones; you can just use any distinct features: up and down, in or out, whatever. As long as these numbers can be represented by particular differences, then computers can be made from transistors, or mechanical components, or even interacting LEGO bricks. If you can represent digits and differences somehow, then a computer can be constructed.

But we can go further, and encapsulate computation in cascades of dominoes crashing together or in flowing water in tubes and bins. If you can cleverly design a way for embodying logic based on how water combines or how lines of dominoes come together—inputs and outputs—or even how LEGO pieces move in a mechanical way, you can build a computer. A computer can be built out of Tinkertoys, and researchers have even shown how to build a primitive piece of computational logic out of crabs. It turns out that swarms of crabs are predictable in certain ways, and this predictability in how they move together can be used to make logic gates, the building blocks of modern machines. I do not, however, recommend running a browser on crabOS, even if theoretically possible.

This is an aspect of a fundamental idea within computer science (we will see more of this when programming languages are explored): that the substrate of computing doesn't really matter. Computation is an idea that is independent of whether we are working with crabs or LEGO bricks or transistors. As long as the something we are working with—whatever it is—can do certain kinds of operations (and it can be even broader than the logic described here), these things are all on some level equivalent.

The reason that we have settled on transistors, and their true power in our case, is that their simple operations are done so quickly and reliably (and in such a small space) that we can take these primitive pieces and combine them into a massive edifice that winks on and off billions of times a second and can be constructed to allow for higher levels of abstraction. We can build computer programs on top of bits of silicon.

THINKING LIKE A PROGRAMMER

But what of these computer programs themselves? What of the nature of code? For many of us, if we've seen code, it appears to be a pseudo-English pidgin, full of words like "while" and "for," along with numbers and symbols, from "+=" to an inordinate number of semicolons or parentheses, depending on the specific programming language. It might look something like this:

```
def factorial(n):
    result = 1
    for i in range(1,n+1):
        result = result * i
    return result
```

I implore you, dear reader, do not look away! Let us dare to confront the nature of code and the mind of the programmer, looking into the sun, as it were, where the fun is.

To write such text requires taking a fuzzy problem and breaking it down into a set of straightforward tasks. But in practice, the mind of the programmer is equal parts logic puzzles, lists of half-remembered function names, and head-pounding frustration at glitches. For writing a program can often mean keeping a huge amount of information in one's mind: how the program is running, what exceptions and edge cases there are that have to be handled, what functions and software libraries can be used, and why something might possibly break. It is fun and exciting but also intensely maddening. Leap years must be accounted for, or temperature conversions, or dealing with names that have unexpected lengths or symbols. All of this is something that has to be accounted for, because computers can be unforgiving.

For example, imagine constructing a thermostat program. Before digital thermostats, they were at their core very simple: a metallic coil and a switch. But even though they weren't computers, they did have a kind of rudimentary intelligence. They adapted to their environment in a specific, and even computational, way. In fact, they were an analog computer of sorts (more on this in later chapters), with a coil that would change with the temperature around it. Very roughly, as the temperature rose, the coil would expand and trigger a switch for cooling, and as the temperature went down, the coil would tighten, ready for the next time it became warm.

Thermostats have become more complex over time. We now have computerized thermostats, for example, and even ones that are incredibly sophisticated and connected to the Internet, both for good (receiving weather data) and for bad (your home might now be hackable via your thermostat).

Independent of the specific mechanism, the underlying logic of a thermostat is one of adaptation to its environment. If a room gets too hot, the thermostat ensures the room is cooled; if it gets too cold, the heat kicks in. Over time, it responds to its surroundings and creates an equilibrium. This process of adaptive response and making sure that the system stays close to a set point is what is known as negative feedback. The output is fed back into the system and makes sure nothing gets too far out of whack (it tries to undo any change, hence "negative" feedback). This is opposed to positive feedback, which is what we normally think of when we hear the term "feedback," like when a microphone magnifies the sounds from its speakers, causing it to become louder and louder, ending up with a squeal. Positive feedback magnifies small phenomena and swamps the system, but negative feedback creates balance.

What would it mean to encode this balancing logic into software? If we wanted to write a computer program that could mimic a thermostat, how would we do this? We would need to know the temperature we wanted the room to be set at, the actual temperature of the room, and a way of altering this temperature by turning heating or cooling on and off. We need if-then logic for sensing conditions, we would need to have some sort of loop that runs this continuously (but also makes sure that it doesn't cause the HVAC system to cycle too often), we might need to store the temperature we want for different days and times, and much more. Embedded within the simple idea of a software-based thermostat is so much: precision, logic, and the capacity to call on both data and abilities from other parts of the machine or even external data sources.

So, when someone tells you that coding is like writing recipes for a machine, he's not wrong. But by no means is that all there is. There is complex recombination, the sharing of inputs and outputs, and the use of powerful and sometimes seemingly esoteric commands. The very act of programming a computer can also teach you more about the problem you are trying to solve; only when you have the solution might you be said to truly understand it. As the titan of computer science and author of the series *The Art of Computer Programming* Donald Knuth has noted, "People often find that the knowledge gained while writing computer programs is far more valuable than the computer's eventual output." Maybe the true treasure was the knowledge we gained along the way.

In addition, computer programs are written as a kind of archaeological tell, built on top of things that have come before them, but which have been buried by time and abstraction. There are similarities to biology here, where evolution works in every eon with what has come before, building from preexisting systems to adapt to an ever-changing environment. When it comes to computation, while I can write a program knowing that physical pieces like transistors are involved, as well as intermediate programs like compilers, I don't have to necessarily think that much about them.

As a result, one of the features of much programming is not just logic and calculation but cobbling bits and pieces together (something evolution does as well). To program involves not just building a new structure but also digging underneath to find snippets that can be useful. Coders chase down functions or programming libraries that can provide what they need and glue them all together. As part of this, many languages offer cavernous programming libraries, each with collections of abilities that either come with the language itself or that other programmers have made available—shelves of magic, all for the taking.

For example, if you want to handle time within a computer program, you don't have to start from scratch. Programming languages make it straightforward to receive the time and operate on it or plop it into a variable; you can simply run a function that tells you the time (abstraction and recombination). There are programming libraries that help with manipulating data, or making graphs, or building neural networks. If you want, for example, to know whether a chess move is valid or to access weather satellite data, you can plug these components together and work wonders.

But relying on other people's work also has a downside. Software can be ephemeral. As the foundation beneath a program shifts—the software packages used to make it run might become obsolete, or the website that provides data changes how it serves information or simply becomes defunct—a program could cease to operate. Think of how many links on older websites don't work anymore, or how many applications on your computer can't run on the new operating system because they haven't been updated. And then imagine that's the story of a programmer's entire professional life. There can be a transience to software, and the Sisyphean acts of maintenance required for long-lasting software to exist are far more incessant than the task of building a bridge or a highway. Software is never a "build and forget it" endeavor. It requires both patience and maintenance.

So, we might try to make analogies to what writing computer code is like, whether to baking, or to building with LEGO bricks, or to composing instructions to a friend on how to drive to your house, or even to gardening. But really, it is not quite any of these things. As the curmudgeonly early computer scientist Edsger Dijkstra—who was responsible for developing foundational computational algorithms—noted, computers and programming have a "radical novelty" to them.

In the past few years, what we think of as programming has begun to change again. Many software developers now work in concert with large AI systems, which help them write boiler-plate code or stitch together different modules and components, completing lines of text for them or even writing entire chunks of a program all at once. These tools have been a boon—looking up the details of functions or how an interface works or even just searching online for how to do some specific weird thing, which is what a lot of computer programming is, is no longer necessary—but they have also made it easier to avoid some of the kind of programming I grew up on.

Nevertheless, programming is still the process of converting one's thoughts into a mechanism for a computer to carry out those ideas and do something. We have come a long way from composing a program entirely in binary or flipping individual switches on early personal computers as a way of controlling its logic. But all of these are part of the programming mindset.

No matter how different things become, and no matter how much has changed in technology—and how much is yet to change—we are in a line from those early days of vacuum tubes and machine code. There is a tale in the Talmud—that ancient multivolume compendium of laws, debate, and stories of the Jewish rabbis—that can provide insight here, a story that involves Moses encountering a rabbi who lived over a thousand years in his future. Ancient time travel aside, the most intriguing part is that when Moses sits in the back of the house of study and listens to this rabbi—Rabbi Akiva—relay his wisdom and learning, Moses becomes entirely confused. Moses, who received the Torah itself from God, does not grasp what was going on. And yet, in the end, how does Rabbi Akiva provide evidence for his insights? He cites them as being derived from a law transmitted to Moses from God. At that point, Moses is calmed: even though he might not understand this rabbi, there is a clear lineage from the old to the new.

Do we learn to program the same way that early programmers did in binary, or to write code the way I learned in high school? No.

I haven't used the programming language C++, which I learned in high school, in about twenty-five years, and I'm not sure I'll ever use it again. But change in both form and technique is part of the nature of any developing field. We revere Isaac Newton for his genius, even though huge numbers of high schoolers are imbibing his then-cutting-edge calculus, and his classical mechanics has been technically eclipsed by the work of Albert Einstein. Similarly, John von Neumann, one of the developers of the fundamental ways for how nearly all modern computers are structured, is said to have noted in 1954 that he didn't see any reason for programming languages beyond writing down zeros and ones: "Why would you want more than machine language?" Things will change, the old guard might have some difficulty adapting, but we are part of a single tradition of computation.

Ultimately, we must understand the history and details of each field in order to understand the frontiers. We can still revere Steve Wozniak's coding prowess for the Apple II, even though no one in their right mind would ever want to build a machine that way any longer.

———

Software is built from cathedrals of code, massive edifices with hundreds of thousands, or millions, or even tens of millions of lines of instructions. This code can be spread across multiple files, worked on by teams of people across years or even decades. They are massive construction projects. Some cathedrals are robust and elegant, others more rambling and teetering than we would like. The Computer History Museum in Mountain View has collected the source code of some of the most seminal programs in computing history, from Microsoft Word to Adobe Photoshop, and these are all valuable, not just in their final output—the programs themselves—but in the abstractions they used and how they were built. I was, for example, surprised to learn that a decent fraction of the Photoshop code was written in assembly—the instructions that control a microchip itself—as opposed to a higher-level language like C.

We might never look at all this source code the same way we now look at untamed wilderness. But no matter what code looks like or even what coding looks like—and it is rapidly changing due to AI—we can also examine how it makes us feel, these aesthetics of algebra and fire.

I am by no means an expert programmer. While I've been programming for about thirty years, I have carefully maintained my amateur status. Nevertheless, when writing computer code, there is this crackling feeling of power: you write a series of lines of text and then execute them, and something actually happens! A line of text appears on the screen, a calculation is made, a complex graphic is elaborated. Your words—precisely composed—have resulted in an action. It's like the scene in *Harry Potter* when Hermione explains the nuances of "Wingardium leviosa," a levitation charm. Text becomes power. And in some ways, this feeling is heightened by the persnickety nature of computer code, by the fact that one misplaced semicolon, or a number value that is one too high or too low, can throw everything off. Because when you do finally get it right, when you can sit back in your chair and exhale with a feeling of satisfaction over having properly assembled the components of computational language, you feel like a wizard of unbelievable powers. While I don't know what it's like to wave a wand or utter a magical spell, I am certain that the feeling of smacking the return key upon code well composed and the feeling of a spell well cast are in the same neighborhood. And that is the subject to which we now turn.

3

Digital Alchemy

Code as Magic

The story of the Tower of Babel occurs in the book of Genesis after the biblical flood and humanity's resettlement of the planet. Humans have chosen to congregate in a single city and have begun to work together—their single language means there are no communication or translation issues—to build a large tower. Confronted with a united humanity involved in the creative use of technology to construct this grand edifice, God confounds their language and disperses everyone. Interpretations of the reasons behind God's actions are many, including the hubris of humanity to challenge God, the fact that people had chosen not to spread across the earth but instead gather in one place, or even a bit of worry from God that people with a united purpose might be difficult to control. No matter. God is unhappy with this single language, used by humans to coerce the world to do their bidding.

The physician and bioethicist Leon Kass has explored, among other things, the morality of human cloning and has also written multiple works on the Bible. Kass has argued that perhaps Babel is actually the world to which we have returned: through mathematics, and now computation, we once again have a single language—code—that can conquer the world around us. Code and computing have the potential to reign over all of creation, providing the abilities of the gods. Computation now appears to be a domain of sorcery and power, godly and wizardlike.

The potency of computation and its similarities to magic can be a lens that allows us to look at the history and future of computing differently. When you take this idea seriously—that computation and magic are similar—you can understand the nature of code and computing better. But seeing how magic clarifies the way programmers build software requires spending time first with the nature, history, and particular properties of magic itself.

THE NATURE OF MAGIC

Magic has a long history with widespread use. While we tend to think of it as an ancient or medieval phenomenon, separate from the "modern" mindset, it is anything but. Magic has coexisted with science, sometimes on a clear continuum with what is considered scientific or true wisdom, such as the pseudosciences of alchemy and astrology. Magic is also deeply connected to the world of religion; in many cases, magic was simply the religion of anyone outside your own culture and society. And while we think of magic as a form of superstition or a childlike approach to the world, in some cases magic was a means—albeit an ineffective one—of grappling with the unknown and even disenchanting the world and making it understandable. This is all to say that what exactly magic consists of is slippery and complicated.

It is nonetheless clear that magic and religion were intertwined with language. Verses from holy texts were used to create amulets with protective properties, and some people ate pages from the Bible to cure illnesses. Psalms were considered to have powers of healing. We now associate the word "abracadabra" with illusionists wearing tuxedoes and pulling rabbits out of top hats, but going back to the Roman era and continuing for millennia after, it was considered a magical word.

Many peoples of the ancient and medieval world compiled entire books of spells to control these linguistic forces that swirled around them. Known as grimoires, these books were found in civilizations around the world and across time, from the ancient Middle East to medieval Europe. And they incorporated the specific ways that letters and words could be combined in order to coerce gods, demons, and other powers of the universe to do one's bidding. Yearning for some divine knowledge? Wondering how to consult a ghost? Want a demon to bring you treasure? Consult a grimoire and see if the necessary spell was contained within it. In fact, since grimoires were focused on the written word, these books were not just repositories of magic but were also thought to have a certain magical quality themselves.

Words, whether spoken or written, could even have a creative power. According to Jewish law, for example, one is required to recite specific blessings before eating different kinds of food, and one of these is translated as, "Blessed are you, Lord our God, ruler of the universe, by whose word all things came to be." Similarly, the Talmud speaks of "letters with which heaven and earth were created." These are words and letters as building blocks of creation.

This view of creation is not unique to traditional Judaism. Christianity, too, recognizes words and their power, and so do many other religious traditions and mythologies. Humanity has long accepted the specific force of language.

The equation of the spoken word with magic was so strong in the world of ancient Israel that one theory for why Moses was denied entrance to the land of Israel by God was that he was punished for trying to use words to coerce the divine. Using words and texts for magic was something that was simply not to be done; it was considered part of pagan ritual and therefore anathema to ancient Judaism.

Magic is also the province of the stories we tell. Specifically, in these tales, words are often granted power. Whether they are written down or spoken out loud, the specific formula of words and texts was thought to have a precise effect in the world. There is "Open sesame" of "Ali Baba and the Forty Thieves," and the precise formulation of Strega Nona (from the children's book *Strega Nona*): "Enough, enough pasta pot, I have my pasta, nice and hot. So simmer down my pot of clay, 'til I'm hungry another day"—plus three kisses blown—ensures that her pasta pot stops generating noodles.

In the world of *The Lord of the Rings*, Gandalf knows many spells to open doors. And a vision of the world of Middle Earth is sung into being by the deities of Tolkien's pantheon, Ilúvatar and the Ainur. My own children used to ask me to look up specific spells from the world of *Harry Potter*.

But it's not that straightforward. There is a craft to magic. You have to spend seven years in Hogwarts to master these skills, or study at a postsecondary school to learn the intricacies of spells. At the university of magic in the novelist Lev Grossman's *The Magicians*, for example, magic is described this way: "Even the simplest spell had to be modified and tweaked and inflected to agree with the time of day, the phase of the moon, the intention and purpose and precise circumstances of its casting, and a hundred other factors."

The book *Babel* by the novelist—and graduate student in East Asian language and literatures—R. F. Kuang employs an intriguing, and highly textual, example of magic, which requires a large amount of education into the imprecise nature of translation. When the same word in two different languages is inscribed on either side of a piece of silver, the magic is the difference between the meanings. For example, pieces of silver are placed in train tracks in the world of the novel with the pair of words "track" in English and "trecken" in Middle Dutch, which also has the meaning "to pull" (this is according to the novel; I can't vouch for this). Into the gap between these words arises the ability for the silver to pull the train cars forward. Or imagine a piece of silver with the word "old" in both English and Chinese. Because the Chinese character apparently has the connotation of durability, use of this silver on machines allows them to become better with age, instead of decaying and breaking down.

The beauty of magic in Kuang's book is not only that it highlights the power and precision of words, but that the ineffable meaning of a text, when translated, inevitably loses a tiny bit of its power. But this power can be brought back into the world through the craft of sorcery. The novel's scholars who are responsible for this magic must be steeped in a great deal of linguistic knowledge, history, and the art and science of translation. Magic, as so many other books depict it, takes a great deal of work and research.

Grimoires, wizardry, and witchcraft—magic is fun to think about, but it is not reality. Nevertheless, while words might not be as magical as people have wished, they still have power, even if it's a power of a very different sort. You can write a book and it can change people's minds. Or you could promulgate a legal code—a code!—and it would prescribe people's behavior. From Martin Luther's Ninety-Five Theses to Hammurabi's Code, language is obviously very influential. And the printing press allowed that influence and power to act at scale.

My son discovered a one-time-use magic word when he was little. Not "please" or "thank you," but "both." It was his first relatively abstract concept. Up until that point, it was nothing but words like "dog" and "ball" and "mama." But when we asked him which of two foods he wanted, and he responded, "Both," my wife and I were so surprised and delighted that he knew this word that his wish was granted. The magic soon wore off, but that word had coerced his parents to do his bidding.

Returning to translation, the reason for its difficulty is that we are providing not just meaningful information when we write or speak but also a certain sense that the words convey. When we choose a particular word, or find the specific phrase we want, or decide to spatter a text with punctuation or emoji, all of these choices can and will influence how our language is perceived. Why else would people perseverate so long over texts and emails? Because the precise formulation of written text will affect those with whom we are communicating. This ability to alter the minds of those around us, even without any magical overtones, is powerful.

MAGICAL CODE

With the advent of digital computers, magic became reality. Beginning several decades ago, you could write strings of letters and numbers and have them operate automatically. Computers allowed words to finally act in the world, in a way that wasn't reliant on the vagaries of humans actually reading a book or understanding a sentence. Computers took away all that messiness, contingency, and ambiguity. And they gave us magic.

Just as magic is powerful but also rigorous, ranging from the precision of text and words required to work a spell, to the nature of esoteric lore, to even the specific ways that these words affect the world around us, a computer program must be precisely written, handling exceptions and strange edge cases. To return to *The Magicians*, magic is "language [that] gets tangled up with the world it describes." Not so far from code, with its precise language and its effects on the world.

This equation of code and magic is not original with me; programmers have long recognized the magical powers that they are working with. System administrators might speak blithely about "daemons" as programs that automatically manage computer processes, and many of us have used a software wizard when setting up a new computer or application. If you flip through *The New Hacker's Dictionary*—a decades-old glossary of computer programming, hacker jargon, and slang, full of often humorous terminology and including words of dubious but insightful usage—you will see the extent to which religion and wizardry have permeated the thinking of programmers. There's deep magic, sorcerer's apprentice mode, magic smoke, magic numbers, spells, and incantations. There's even a particular programming manual that is referred to as the "Old Testament" and a type of program known as a dragon. It's as if writing software was something done only when these coders were taking a short break from reading about wizards or rolling twenty-sided dice.

The journalist Clive Thompson, in his book *Coders*, describes one of the co-creators of the computer game *Myst*, an evangelical Christian who would periodically break from writing computer code to note—cribbing from the book of Genesis—"It is good." Thompson also notes that one of the functions found in Perl is called "bless." There have even been tech start-ups named Spell and Potion, and one of the largest tech companies in the world is named Oracle.

Fred Brooks, a computer scientist who helped design systems at IBM, wrote the seminal work *The Mythical Man-Month* on software engineering in 1975. He made this same point in that book: "The magic of myth and legend has come true in our time. One types the correct incantation on a keyboard, and a display screen comes to life, showing things that never were nor could be." Ted Nelson, a computer theorist of Brooks's era who coined the term "hypertext," in his book *Computer Lib/Dream Machines* uses magical metaphors to think about computation as well (though sometimes only to disabuse readers of these metaphors), from "The Magic of Data" to "The Magic of the Computer Program." The influential computer science textbook *Structure and Interpretation of Computer Programs* makes an analogy between the novice programmer and a sorcerer's apprentice. The term "magician" in its index even refers you to the term "numerical analyst." And one of the stories deeply embedded within hacker lore is known as "A Story About 'Magic,'" telling a tale from decades ago of an unexplainable switch labeled "magic" that, when flipped, would somehow cause the entire machine to fail.

We can also see this association of code and magic in fiction.

Science fiction and fantasy are the worlds that many coders swim in. Science fiction stories show us imagined technologies, inspiring scientists and engineers. But fantasy provides hints for how code should crackle. Write some text and coerce the world. Utter words and make something happen. When I look at source code, I should see the glamour of its magic-laden nature. Intriguingly, fantasy and science fiction recognize this, sometimes creating a convergence exactly at the point at which magic and computing intersect. Fantasy and science fiction blend together when it comes to computer code.

"Seventy-Two Letters," for example, is a short story by the famed science fiction author Ted Chiang set in the Victorian era, and it plays in the fantasy-meets-science-fiction-meets-computing space. In the story, Chiang elaborates an intriguing combination of kabbalistic manipulation of Hebrew letters, robot programming, the animation of golems via these collections of letters (something to which we will return), and even the biological language of genetics. The online writer Scott Alexander's novel *Unsong* playfully explores this kabbalistic manipulation as well, its power and rigorous nature, and even how these manipulations could be protected by intellectual property law.

The recognition of the similarities between magic and computation is not just intriguing and fun. By taking this analogy seriously and trying to see where it works, we can better understand computation and code. In particular, we can do this by looking at four areas where they intersect: names, the kabbalistic recombination of letters and words, grimoires, and sacrifice.

NAMES

A name is something profound. According to many traditions, to know the name of a thing, a creature, or a person is to give yourself power over it. In the ancient world and in stories, the names of gods and demons were thought to be replete with this power. In fact, ancient Judaism felt the four-letter name of God—YHWH—was so holy that it underwent a process of distancing. At one point, it is said that the only time this tetragrammaton was uttered was during the Yom Kippur service in the ancient temple by the high priest, and eventually it ceased to be spoken aloud entirely.

There is also an oft-told etymological tale that the word for "bear" in old German languages is actually the word for "brown one," because the original word was thought to bring forth bears every time it was used. If you didn't want to get mauled by a bear every time you spoke about one, the thinking went, perhaps you should use some sort of euphemism instead. Some people have even claimed that not only was the original word not used but it was lost, because the euphemism was used so often. It wasn't: the original unsayable word is related to the word "arctic." Nevertheless, the ability to refer to a thing by a name was imbued with a great force.

We see this in our stories as well. The fairy tale of "Rumpelstiltskin" hinges on whether the main character knows this imp's name. Know his name and you defeat this creepy dude. Similarly, in the Earthsea novels of Ursula K. Le Guin, the magical system of this world revolves around the idea of true names for magic. As per Le Guin: "For magic consists in this, the true naming of a thing." Of course, learning and determining the true name is a hard and complex task, and fractal in its complexity—recall the importance of training and long years of study in the practice of magic.

In computers, one of the clearest instances of a true name is that of variables. When you write a computer program, a variable is a type of placeholder. It is a name that refers to a location in memory where information is stored. For example, in Python, if I want to store some number, I can simply declare a variable this way:

```
someNumber = 137
```

Now this variable "someNumber" has the value of 137. If I want to change it or modify it, I can easily do so, such as by adding a number to itself (this makes it 147):

```
someNumber = someNumber + 10
```

While the details of how all of this works vary from language to language, variables are a kind of true name. Know the particulars of a variable, and you can modify the software's behavior. While this sounds straightforward, variables can often have deep and complicated structures within them. You might have a variable that contains a list of objects, some of which are also lists themselves, nesting items within items. There are variables that act as dictionaries, or are matrices of numbers, or hold huge chunks of text. Identifying an item of interest within a variable can be far from a trivial endeavor. True names take work and time. As a result, there is an art to naming a variable in your code.

For example, don't give a variable a simple or common name unless it will be only used briefly and quickly evaporate. When particular bits of memory that a programmer wants to keep separate end up having the same name, the computer will view them as the same thing, and there's no telling what might happen. For example, generally don't name a variable that is holding a number n. Even though that seems like a straightforward and appropriate name, there is a good chance that some other variable will also be called n, the program will combine these in unexpected ways, and bugs and glitches will ensue.

So, too, should a variable be at least somewhat descriptive. Don't give a variable a sort of vague placeholder name, only to realize at a later point that you are not entirely sure what it contains (someNumber is very much not a good name). A programmer will look at one of these generic variable names and simply have no idea what it's supposed to be doing, even if he himself wrote the code several days before. And if the programmer doesn't know exactly what is going on, he must either spend an inordinate amount of time trying to figure that out or just think he knows (but doesn't quite) and cause more issues.

As a program grows, the programmer's task—figuring out what variables contain, what they are meant to do, and how they operate, and then ensuring that nothing overlaps in unexpected ways—grows increasingly complex. Into all of this have stepped naming conventions: rules behind how variables should be named, a sort of manual for spellwork. There's a style known as CamelCase, which involves giving variable names a multi-word phrase that is all combined into a WordThatHasCapitalizationButNoSpaces. Another acceptable alternative is this_sort_of_variable_name, though different programming languages have different conventions, sometimes for the different types of variables themselves. The kind of name can give you a sense of a variable's power.

But the most powerful variables might be global variables. These are Sauron's rings of power in a land of plain gold wedding bands.

All variables have what are known as "scope," the realm of computational space where they can be referenced. For example, if you create a variable within a function, it is only able to be used when that function is run. And once that function has done its thing, the variable and its contents are lost to the textual ether. You have gone outside the scope of that variable, and the variable's name is no longer viable. For example, once I'm done with some loop, the variable that acts as a counter for how many times it's being run vanishes.

The extreme version of this is a global variable: its scope is the entire program. Such a variable can be examined and modified from any module or function at any point in the running of the software after it's been declared. No doubt there are situations when this is useful—just as knowing the rules of grammar allows a writer to bend them when necessary—but overall, global variables are considered incredibly poor form. What if this variable has a common name that is used by another variable elsewhere in the code? Or what if you write a line that modifies this global variable, not accounting for all of the later effects it might have? Or, if there are multiple people working on the code, what if they have done something to the global variable that you didn't anticipate? A global variable is a true name with far too much power. This is sorcery of a magnitude that no programmer should be granted. Hence, the general rule is to avoid global variables as much as possible.

The converse of a true name in code might be a magic number. Magic numbers are specific numeric values that are vital for the function of a piece of software, though their specific values might very well be inscrutable. These are numbers that are written directly into computer programs, rather than defining a constant variable, which can be modified if needed.

A good example of a magic number—specifically one that seems uninterpretable—is from the gaming engine developed for the game *Quake III Arena*. Buried in the computer code is this line:

```
i = 0x5f3759df - ( i >> 1 );
```

Insightfully annotated with the comment "What the fuck?" this code is bizarre, in that there is a strange and mysterious number, 0x5f3759df, that somehow helps a procedure rapidly calculate an inverse square root (1 over a number's square root). It turns out that there is a lot of mathematics behind why this hexadecimal value is so important, involving a particular mathematical method to quickly converge on a good approximation of the answer. But rather than the programmer describing how all this is done, there's just a weird number in the code. And it works, somehow tapping into the eerie strangeness of the world of mathematics. It's the deep magic of a precise value—perhaps a convoluted name of sorts—that allows everything to work.

Whether magic numbers or variables, names are something not just in magic but in code as well. And recognizing this can give us insight into how software is built.

We now turn to a different aspect of magic found in programming: the recombination and manipulation of text.

COMBINING AND MANIPULATING TEXT

In addition to the magic of names, there is the strength of combining and ordering specific sequences of letters. Probably the best-known example of this is in the Kabbalah. The kabbalistic tradition grew out of medieval Jewry, and one of its important texts is the ancient *Sefer Yetzirah* (the Book of Creation, or Book of Formation), which discusses the details of using the letters of Hebrew for creative powers. This Book of Creation explores the details of the Hebrew language, and how its twenty-two letters can be used to make everything. The results were medieval kabbalistic rituals for taking these letters and arranging them in certain ways to act as magical formulas, or even for programming a golem, a claylike automaton that operates based on the specific string of letters used to grant it power.

We don't have Kabbalah in computing, though we do have short strings of code that feel magical. The closest we might have to this alphabetical manipulation rife with power is something known as a regular expression.

Regular expressions are a simple programming technique that involve writing cryptic strings of symbols that are able to match specific categories of text. Searching for a specific word or phrase in a text is straightforward, but what if you want to find any time a date is written out? Or every spelling for the word "Chanukah," of which there are far too many? Or all forms of HTML tags in the source of a web page? Regular expressions allow you to write patterns that can extract every form of these bits of text. For example, `colou?r` will match both "color" and its British variant "colour," because the question mark means zero or one instance of the character before it (so either one or no *u*). Or becoming sophisticated, here's one that is designed to match dates written in "mm/dd/yyyy" form:

```
^(0[1-9]|1[012])[-/.](0[1-9]|[12][0-9]|3[01])[-
/.](19|20)\d\d$
```

No longer must you manually scan a document for each instance that a date is written out. Run this once, and it finds them all simultaneously. Regular expressions—these precise kabbalistic strings—are the "Accio broomstick!" of text processing.

Another example of such textual power is found in the telegraphic instructions of the command line. The command line of the Unix operating system is a kind of computational servant: a text prompt that waits for you to type a bit of text, hit the return key, and then watch your commands be followed, expertly and literally. Unix, due to its fluency at connecting programs together, allows you to create a series of commands and then string them all together, like the pieces of a Rube Goldberg machine, with the output sometimes being far more sophisticated than you might have imagined.

For example, the primitives of Unix are rich and open-ended enough that you can straightforwardly construct a spell-check function without one ever having been programmed into the machine. You can do it by typing the following into the command line, a classic example developed decades ago:

```
cat document | tr A-Z a-z | tr -d ',.:;()?!' | tr ' '
                '\n' | sort | uniq | comm -1—dict
```

What does this unearthly string of letters mean? Without going into too many details, it takes a file, makes all text within it lowercase, eliminates punctuation, sorts the words and eliminates duplicates, and searches for them in a dictionary. The power of spell-check is conjured out of textual manipulation commands.

Whether it's the textual combinations of the command line or the complex ways that regular expressions can be generated, these are close to the kabbalistic manipulations of letters and text.

GRIMOIRES

The grimoire—a collection of practical magic, with spells and charms compiled into a single book—is another example of magic found in the world of computing. We can see an example of the grimoire in computing by examining a text from the early days of computers: *A Million Random Digits with 100,000 Normal Deviates*. Published in 1955 by the RAND Corporation—founded to provide the intellectual underpinnings for the deep intertwining of the industrial and the military—it is exactly what you might expect, containing column after column, for hundreds of pages, of provably random numbers. These million numbers were generated in a most rigorous manner, the RAND Corporation assured the reader, using a sort of "electronic roulette wheel" based on a large number of electrical pulses. A sampling included such gripping sequences as "14556 82726 34676." The researchers, squirreled away at RAND, had trapped the primordial effluence of the gods of chaos between two covers.

But what was the purpose of this unfiltered chance? In humanity's rush to civilize its surroundings and subdue nature, could there possibly be any sort of market for randomness? The answer is an emphatic yes. These numbers could be the raw materials for any algorithmic needs that might require moving beyond deterministic rules and formulas. Randomness is found deep within such matters as code making and code breaking, scientific experimental design, simulation of the real and messy world, and simply the many, many ways in which something might deviate from random chance. You can only know how unexpected something is if you know the true shape of randomness.

A Million Random Digits with 100,000 Normal Deviates is essentially a grimoire: it provided the raw materials for a whole slew of algorithms. While we no longer need RAND's *Million Random Digits*—computers can now generate randomness, or at least a simulacrum of randomness—there are other grimoires in the realm of computing. One book that provides mechanisms for generating this kind of pseudo-randomness is *Numerical Recipes*. This text collects numerous useful algorithms for a wide variety of tasks in addition to generating randomness, from finding roots of equations to manipulating matrices of numbers, and is certainly a grimoire for computing.

Another computational grimoire is HAKMEM, a 1972 memo from the MIT AI Lab that along with lots of mathematics includes hacks and programming techniques compiled by those who had gone through the lab over the years (and to which we will return in one of the final chapters). There are insights into computer graphics, ways of manipulating a machine's memory, and how to calculate pi. It's a manual that leans into its cryptic nature.

And there is Stack Overflow, the website for an ever-growing repository of programming, a place to share spells and bits of lore. It is an online forum for asking for and receiving information on programming tasks, from basic algorithms like flattening a complex list of items to how to handle various compilation errors. It is a living document of the collective wisdom and ignorance of software development, where a programmer can go to find out how to use a software package properly or write a particular function. It is the grimoire to end all grimoires.

SACRIFICE

Another aspect of magic and ritual is that of sacrifice: something of value must be given in order for the spell to work. But if, in our stories, a demon requires blood or the death of an innocent for its curse to take effect, what is the equivalent in the world of computation?

The realm of artificial intelligence and machine learning provides an example. Do you want a powerful system that can generate text, or code, or images? Then you must place a large corpus of humanity's creations upon its altar. These systems are greedy demons, insatiably consuming vast quantities of data. And in their automated output, you can see hints of what has been sacrificed: Getty watermarks on AI images, stylistic mimicry of artists, slightly modified copyrighted computer code generated by a powerful auto-complete. The creative work of artists, photographers, and programmers has clearly been devoured to generate these outputs.

Many creators are disturbed—rightly so—by what is required for these AI systems to work, the creative sacrifices that are incinerated by these massive data-processing centers in order to give us the wonders of artificial intelligence.

But these systems only work with sacrifice. The sophisticated AI that we are familiar with can only learn by cranking through huge amounts of data, sifting through these textual corpora or towers of images to extract patterns. AI requires a huge number of examples of cats to extract "catness," or Wikipedias worth of writing to predict the next hundred words of a chunk of text. In this way, at least right now, these systems are different from people, who learn much more rapidly and without nearly as many examples. So the power of AI comes at a price. There are vast resources that must be consumed to train these systems (computational power and electricity), and these systems stand upon a heap of creative work of human beings. As a result, many people involved are now asking whether creators should be compensated for their work if it's included in the training data, or whether people can opt out of having their information included in these datasets entirely. In other words, should these sacrifices only be given willingly?

AI aside, computing is not free. You might think of the cloud or the Internet as some intangible thing that exists outside the material realm, but computation is a very real and physical system. There are colossal data centers hosted across the globe—to make it easier to access your information wherever you might be using a computer—and cables crisscrossing the planet to pass information back and forth. Computational processing is a real resource, with programmers requiring access to electricity and chips that can churn through data. Bitcoin is close to a pure example of such cost and sacrifice: computational resources are the cost of mining this digital currency. Much as the celestial beings of Middle Earth elucidated a vision of their world through song, perhaps this is nothing more than machines singing money into existence, spinning code and computation into value.

———

Technology has always been a part of magic. Whether it was the introduction of papyrus or the use of the printing press in making grimoires, novel technologies are grist for the world of magic. The very act of printing was once viewed with suspicion; in the folklore of the Renaissance, a devil was said to exist within every print shop. When mobile phones became common in Nigeria, there was a period around 2004 where there was a widespread belief that answering calls from certain numbers—which were even printed in newspapers—could lead the user to instantly die or go mad.

Do not be surprised, then, to learn that some have taken the analogy of computation and magic too far and have even tried to do "real" magic with it—the weird and nonscientific kookiness that I have avoided exploring until now. These are the folks who think that the world of magic is real and try to generate powerful sigils—symbols and patterns capable of magic—via software. That's . . . interesting, I guess. But this is very different from what I am exploring here.

How people relate to artificial intelligence in particular—something to which we will return—also requires increased attention in relation to magic, as the artificial and the autonomous are increasingly getting all intertwined and confused. With the advent of artificial intelligence, we are beginning to see new analogies to magic, particularly in the way we write prompts, the sentences or phrases that activate an AI system to create an image or text. You can ask an image generator for a hair-band rock album featuring the Muppets, or for world's fair posters from medieval times or hosted on the moon or Mars. But as many people who have played more than a bit with these systems know, there are gaps between the image in one's mind—or the poem of one's dreams—and what the machine will give you. Which means that you have to learn the temperament of each system, which additional keywords to whisper to these tools—such as "photorealistic" or "vaporwave"—in order to yield the desired results. This is spellcasting, plain and simple.

———

For nearly two thousand years, it is said that rabbis have explored the combinatorial powers of the Hebrew alphabet, mixing letters together in the proper order for the purposes of creation. There is a story from the Talmud of two rabbis who would study these techniques every Friday, create for themselves a calf from nothing, and then eat it. This is related to traditions connected to Kabbalah and tales of golems, those creatures that were imbued with autonomy by virtue of this magical technology. These tales include the story of a clay golem that, when it became monstrous, had the name of God removed from it, in order to halt its destructive powers. The most common version of this story is attributed to Rabbi Judah Loew, also known as the Maharal of Prague, who lived during the sixteenth and early seventeenth centuries.

Intriguingly, three giants of computer science and artificial intelligence all have within their family's lore the tradition that they are direct descendants of the Maharal, this rabbinic sorcerer said to be versed in the ways of textual manipulation for the purposes of creating life and intelligence. The scientists, Gerald Sussman, Marvin Minsky, and Joel Moses, all worked at MIT and were closely involved in the origins of these fields. Sussman's grandfather apparently even passed along the specific utterance—which Sussman immediately forgot—needed to reanimate this golem said to be stored in the attic in a Prague synagogue.

Of course, this story might be best viewed as a fable; although Sussman shared this story, even he does not believe that this tradition is true and he thinks that golems are only silly tales. But the fable's power lies in how it ties these titans of AI genealogically to kabbalistic magic, showing a more direct line from magicians to software developers than we might have recognized.

———

Language is a powerful material. Prosaic words and phrases, even without all the wands or the whirlwinds or the wizardry, are magical enough. We argue, I lay out my points, and you are convinced. Your mind has been altered by a stream of words, and you might even act differently now (the devil is in the details, of course; we are still subject to all those unresolved bug reports known as cognitive biases).

That being said, humanity—deep in its subconscious core—seems to have always yearned for more. And so language became the substrate for the more potent stuff of magic. But only with the advent of digital machines did something truly change.

There is an early digital computer with the name Golem, something that I myself have seen. It was located in the basement of the computer science building at the Weizmann Institute in Israel, and I remember stumbling upon it during a summer I spent at

Weizmann when in graduate school. It's a large, yellow mainframe machine, tucked near a stairwell, according to my memory, and was astonishing enough that I recall taking a photograph of it (which I sadly no longer can find).

We have come a long way from the Golem machine and its bulky harvest-gold form. I can now write computer programs on my phone, a device with millions of times more processing power than the machines that guided humans to the moon. The experience of programming a machine right now is something quite specific: it's an exercise in text and not in frustrating strings of binary, and we are often able to ignore the physical details of the computer. Furthermore, code is still persnickety enough that it requires the complexity and detailed-oriented nature of a programmer, as opposed to a natural conversation with an AI, for example. In other words, our current window in time might in fact be the ideal environment for thinking of code as magic.

The early days of writing out lines of zeros and ones might have been too mind-numbing to feel quite as magical, and our potential future of simply requesting a computer program might just feel too easy. Using text to program—but the complicated and difficult text of our current programming languages—might feel the most magical of all. Nevertheless, whether it is if statements and variable names or more natural words and phrases that are the substrate of magical power, computing is still the premier realm where the language of precise and powerful magic happens. The advent of computing is the key feature; all the rest is detail.

"Antediluvian" refers to the era before the biblical flood and "prelapsarian" to the brief, idyllic period before the expulsion from Eden. I do not know if there is a word quite like these for the era before the confusion and dispersion from Babel, but it seems we have returned to this world. Despite the proliferation of different computer languages and computational methods for interacting with these machines, we have returned to a single language that commingles power, magic, and communication: the world of computation.

It's now time to see how we build software with this powerful language.

4

Out of the Whirlwind

Open Source Software and Open-Endedness

B ack in the day—in the 1970s and 1980s—people reading computer magazines would often come across pages of source code that could be entered into their machines. Want to play a game or draw some cool shapes and patterns? Here are pages of computer instructions. You would spend some time entering them in by hand—which, in the process, presumably gave you some insight into how the software actually worked—and you could play with it as much as desired. It was freely copyable computer code that could be turned into action.

Without the underlying code, a program is an impenetrable nest of machine instructions, bits and bytes of commands to the computer that are almost impossible to decipher. But with the code, not only can it be run, but it can be inspected, modified, and truly made one's own.

Once computer code is available, it can be shared and recombined. But even more than that, this open source approach allows for a rich kind of textual tradition—software developers passing down texts and information, modifying and combining these over time—in the world of computing.

After our examination in previous chapters of the nature of code itself, it's time to move to the next level and begin thinking about how software is built: how its textual tradition developed and how it allows for software—these objects built from the magical medium of code—to grow and change over time and build upon itself. Because when this happens, a long-lived and open-ended body of knowledge is created, one that allows for a supernova of software growth.

———

The sharing of source code was the way that much early software was distributed: not as a final, runnable program specific to whatever machine type it was developed on but as the raw code. The source code could contain the higher-level logic of how a program worked.

An early example, developed even prior to personal computers, was the computer program *Spacewar!*, generally considered to be one of the first computer games. It was developed by the early hackers at MIT, including one whose nickname was Slug, and even led to the first joysticks being cobbled together specifically to play the game. *Spacewar!* has many of the features of a modern game: real-time graphics, multiple players, and the requirement that the players have fast reflexes. Players control one of two spacecraft, slingshot around a central star, and shoot torpedoes, trying to destroy each other. There is even the ability to jump—via "hyperspace"— to a different, random part of the screen. All these features made it surprisingly playable, on par with the modern video-game world. My son and I had a great time playing a version of *Spacewar!* at the Saint Louis Science Center game exhibit and we have periodically played an emulated version on my computer since then.

Spacewar! was written in assembly language and designed for the PDP-1, a large digital minicomputer (a computer about the size of a refrigerator). But *Spacewar!* wasn't sold. Its code filled twenty-seven pages, and the source text was shared widely, allowing it to run on PDP machines everywhere. *Spacewar!* also changed. It mutated and evolved as it spread, as people developed ideas for certain features or fun concepts to add to the game. Eventually, it was modified to run on new and different types of computers, and different details and abilities were added.

But *Spacewar!* was not alone in this way. Many early computer games were similarly shared affairs, copied over and over, including the classic *Colossal Cave Adventure* and a text-based *Star Trek* game.

These features of sharing, modification, and community are therefore not just utilitarian means of distributing applications but mechanisms for combinatorial development, which allow exciting and new types of programs to be made. And they are all hallmarks of the idea of open source software. We can see the power of these features in the relationship between abstraction and open source software.

FROM ABSTRACTION TO OPEN SOURCE

We explored abstraction earlier through the idea of building blocks that can be used to construct more complex technologies. But it's worth spending some time on how the process of sharing and combining—based on abstraction and reuse—allows for a never-ending explosion of software complexity.

There is a set of primitive building blocks in computing that we take for granted. Almost like axioms in a mathematical proof, we don't have to think about the details of checkboxes, buttons, windows, or whatever has been granted to us by abstraction. When Apple first released the Macintosh, software developers who wished to build applications for it were able to consult documentation known as *Inside Macintosh*, which explained the framework for generating windows and buttons all in the style of the original Macintosh. There was no need to reinvent them.

Instantly, anyone who wanted to make a Macintosh application could use these primitives to build software that felt consistent with the Macintosh aesthetic. But while this allows for a software uniformity, if we really wish for a combinatorial open-endedness—allowing novel and surprising software to be easily created—we also need to be able to develop new primitives.

To return to the mathematical analogy, mathematicians have a variety of pieces they can work with when proving a mathematical statement. There are the rules of logical deduction and inductive reasoning, as well as a set of axioms that are available to them. These are the statements so basic that we take them as given and do not prove them. But we actually do not have to start from scratch every time a proof is being made, with just this small set of parts. Mathematicians are also able to work with a preexisting set of theorems: previously proven statements that prior mathematicians have labored over.

The set of axioms and theorems can be added to over time by other mathematicians, creating a whole set of tools that can be applied to whatever a mathematician might wish to prove. In the same way, software developers want to be able to add to and modify the set of primitives and pieces available. Abstraction is sort of like that. Once I create a piece of code that does something, I don't have to peek under the hood. (For the most part. We will see some of these exceptions as we explore the computational cosmos later.) The code just does what we need, the same way a theorem does. It doesn't need to be re-derived every time we want to use it. And we can build with it. These are the primitive building blocks of text—components of an ever-growing tradition—that we can use to make better software.

Open source software takes this idea of abstraction and turns it into a process that oozes across the boundaries of a team or organization. An individual can abstract away the details in her own program, of course. For example, I've made a computational object that generates a bouncing ball; now I can just refer to the ball without rewriting the details over and over. But open source goes beyond that, past even the abstractions of one team in a large corporation that can be used elsewhere. Everyone now has access to these abstractions.

We can move beyond buttons and windows to larger components. The components of a text editor, for example, are now available as an open source primitive—courtesy of other programmers—for anyone to snap into their computer program. Another primitive emerging is the infinite canvas: where you can move around a never-ending landscape in your design or drawing tool, with what is visible on your screen being but a small fraction. As the number of primitives grows, each can be combined into more and more components, which might become primitives themselves.

Software developers don't have to start from scratch; they can grab off-the-shelf components and concentrate on the novel piece of software they want to build, rather than reinventing each piece.

This idea of composable modules is a powerful feature of software in general. But when it can occur entirely independently of any team or individual—the people making these modules are different from the ones using them—then there is the potential for a combinatorial explosion of software. When my son is given a new type of LEGO brick, it provides him with a new way of building with all the bricks he already has. Anyone can use these pieces and play with them however they want. This is the power of open-ended recombination in software.

But of course, there is more to open source than just this open-endedness. There is the collaborative aspect of software development, too, where anyone can theoretically contribute to developing or modifying a piece of software. When anyone can be a part, there is the hope that bugs and glitches will be stamped out more easily, because a far greater number of individuals will be involved in finding and fixing them. And there is also the fact that this software is publicly available and no one has to pay for it: it somehow bubbles up from the collective computational unconscious and can be copied onto your machine.

Beginning in the 1980s, free software and open source software became an identifiable ideology. Or really, plural ideologies: While the terms "free" and "open source" might be roughly interchangeable, the free software movement is very much countercultural and focused on the freedom of the user. The open source movement is a bit softer around the edges and is focused more on the developer and on arguing the numerous benefits of software that can be shared and built upon.

But before all of this—before open source was even a twinkle in its creators' eyes, before shareware was being swapped on bulletin board systems—there was Unix. And examining Unix will help us to better understand the nature of software as a tradition of textual evolution.

What is Unix? It is an operating system framework that can be run from the command line, managing files and folders, handling memory, controlling the display, opening programs, and more. However, Unix has one additional key feature. It is no longer proprietary. Initially developed in 1969 and first released in 1971, it began as the property of AT&T. This was the time when Bell Labs—AT&T's research and development organization that was chock-full of some of the best scientists in the world—still had a broad mandate and could be involved in matters of astonishingly wide usefulness, from inventing transistors to detecting evidence of the big bang (both of which won Nobel Prizes for Bell Labs). In addition, some of its researchers created Unix.

Based on ideas for a previous operating system, Unix was initially built over the course of a few weeks. The lore, and it's a true story, is that Bell Labs computer scientist Ken Thompson's wife had gone on a trip with their young son for three weeks, and Thompson used this time to develop the entire framework of Unix. Given that modern operating systems can require entire teams to build, this level of productivity is something that my mind shudders to grasp.

Originally written in assembly code for the DEC PDP-7—one of those digital computers found in universities and corporations—Unix was slowly added to and then converted into the programming language of C (this language was also developed at Bell Labs). Once Unix was written in this higher-level language that did not have the same level of machine dependence, it could be more easily run on a whole variety of computer types. In other words, it had become "portable" and could be transferred from one machine to another. Alongside becoming machine agnostic, Unix could be added to over time, accreting a whole host of tools with names that might as well be rejected *Star Trek* alien species: grep, yacc, lex, cat.

However, because AT&T was a government-regulated monopoly, it had certain restrictions. Among them was the stipulation that Bell Labs give away Unix to any university that asked. To be clear, this was far from open source; if you were a corporation, you had to pay a hefty license fee to access Unix (for universities, it was much cheaper). But over time, Unix spread across the country, and it eventually spawned entirely novel rewrites of the operating system. These new versions used the same sort of system calls that Unix did, but under the hood, the computer code was entirely independent. And once independent, it was no longer subject to the intellectual property restrictions of AT&T. The two main branches that formed from this were the Berkeley Software Distribution (BSD) and Linux. Linux was the first truly open source version of Unix, in the sense that it was both available to all who wished and built in an entirely collaborative way. And from these two branches, a Cambrian explosion of Unix flavors occurred, from Solaris and FreeBSD to Xenix and Darwin (the last one being the variant that exists deep within the bowels of modern Mac computers). Many of these were truly open source, able to be shared and modified over time.

The Unix operating system had become a kind of evolutionary text, a tradition that generations of programmers were part of, adding to it and modifying it over time. It is the ur–operating system of the computing era.

UNIX MYTHOLOGY

Neal Stephenson is known for writing generally brick-size science fiction novels. They range from cyberpunk to historical fiction and are among my favorite books. But Stephenson has also dabbled in nonfiction. He's written about undersea fiber-optic wires and about treadmill desks. And he also wrote an essay that was turned into a short book called *In the Beginning . . . Was the Command Line.* It was written in the late 1990s, so it is very much of its time. For example, there is an astonishing amount about the Be operating system, something that, if you were not playing in the Macintosh and Mac-adjacent world at that time, you likely have never heard of. (In short, the makers of Be were hoping that it would be bought by Apple and become the basis for the next version of its operating system. Instead, Apple purchased NeXT, Steve Jobs again became the CEO of Apple, and this little bit of history was forgotten by basically anyone who is unfamiliar with the line of blinking lights on the BeBoxes. But I digress.)

Nevertheless, within Stephenson's short manifesto devoted to praising the command line and operating systems like Linux—though there is a lot more in this delightfully rambling essay—he compares a number of different operating systems using a strange but delightful metaphor: vehicles. There's the Macintosh operating system, which is a beautiful, sealed European sedan—impossible to fix or tinker with but, wow, is it cool. There's Windows 95, which is some sort of ugly station wagon (remember, this is a polemic!); BeOS, which is a Batmobile; and Linux, which is a formidable and tricked-out tank, free for the taking and ready to drive.

Whatever you might feel about these analogies, Linux and Unix are unbelievably powerful, infinitely malleable, and just *there*, like the air we breathe, or a star in the celestial sphere. They have fundamental ideas around hierarchical files and folders, specialized languages for specific use cases, and the ability to connect programs together in a natural and powerful manner. Stephenson even compares Unix to the ancient *Epic of Gilgamesh*, a comparison to which we will return, but that undervalues Unix's sheer antiquity.

Unix is old. Using some very rough estimations and analogies, we can see how much of a graybeard Unix is. If digital computers first came onto the scene around 1945—making this technology only about eighty years old—then Unix arose about a third of the way into all of computing history. And yet it is still being used. Even though we have moved on to newer programming languages and from mainframe machines to personal computers and even to phones that slip into our pockets, Unix persists. It is found in refrigerators and undersea devices. It is everywhere.

If modern humans appeared about three hundred thousand years ago, then Unix would not be the *Epic of Gilgamesh* or the Bible, which are babies in comparison. Even cave paintings are only several tens of thousands of years old. If we run the numbers, Unix would be an ancient technology or text that is about two hundred thousand years old and yet still widely being used. Whether this shows a certain amount of stagnation on the part of computational innovation or the sheer pliable versatility of Unix—and an almost atomic necessity and universality—Unix is still the best that we have.

While Unix came out of the gate with a great number of features, this first version is not truly what we use today. And that is because the source code (though originally a trade secret) and its functionality were available to all. Due to that, Unix *became* a great piece of software. It took years of honing and development for it to become what we have now.

Most open source creators hope for their work to be used widely and for it to have staying power. But in reality, open source software is a poem-filled bottle thrown onto the turbulent ocean of code: the odds that anyone will find it, read it, and then make it their own, adding to its beauty and making it more powerful, are vanishingly small. Too often, open source projects are more like message-wrapped rocks tossed into the sea; they sink without a trace. But if that message is indeed found, pored over with care, and adapted to one's personal and special needs, then something wonderful can happen: its text has staying power.

The comparison most apt to open source, then, is not rocks or bottles or anything ocean related, but to the stories that we tell as societies. And perhaps the clearest example is that of ancient myths.

The mythology of the ancient Greeks infused their society, from their rituals to their entertainment. But these stories had no single, canonical version. Instead, they were told and retold over time, with continuous borrowing and adaptation. In *Gods and Mortals*, the classics scholar Sarah Iles Johnston examines this in more detail, providing lots of examples, and concluding that "it was in this spirit of both tradition and constant innovation that the Greeks told the same myths for more than a millennium." These ancient Greeks were full-on remixing their civilizational public domain.

Over time, though, each ever-retold story was sanded into a gem of a tale that could last all those years, not necessarily because the written text was preserved—to be found in a clay pot in some desert cave millennia hence, or etched into a rock so it could not be forgotten—but because there was a community devoted to its recounting.

This balance between tradition and innovation is found in other ancient cultures as well. In Jewish law, for example, the stories, debates, and decisions are continuously added to and modified over time. In particular, before the Talmud, there was a strong tradition of oral transmission. But whether written down or not, ancient stories, texts, and laws were passed down from generation to generation, and as long as they were modified, remarked upon, debated, and discussed, they became living texts.

Open source has similar features. In his book, Neal Stephenson draws parallels between the *Epic of Gilgamesh* and Unix:

> Unix, by contrast, is not so much a product as it is a painstakingly compiled oral history of the hacker subculture. It is our *Gilgamesh* epic.
>
> What made old epics like *Gilgamesh* so powerful and so long-lived was that they were living bodies of narrative that many people knew by heart, and told over and over again—making their own personal embellishments whenever it struck their fancy. The bad embellishments were shouted down, the good ones picked up by others, polished, improved, and, over time, incorporated into the story. Likewise, Unix is known, loved, and understood by so many hackers that it can be re-created from scratch whenever someone needs it.

But this long tradition of maintenance and change isn't always a given when it comes to code. Code requires a community of users, maintainers, and contributors in order to be a truly living thing. Without this, it will wither away, perhaps to be rediscovered by some computational archaeologist. But it will be no more than a program preserved in amber.

These two conditions have been described as either active or static states, with something like Linux in the former category and the source code for Apollo 11's guidance computer in the latter. One has a living, breathing community. The other can be read, looked at, and examined, but more for posterity than anything else.

Of course, moving from a static to an active state incurs costs, such as the need for maintenance and interoperability, but those are the trade-offs of a living text. Only when a text is alive can it be part of a conversation of ideas and stories. This is why there is such a lively discussion around ensuring that creative works can become part of the public domain. Otherwise they are the province of a single individual or corporation (or even worse, are just locked up but no one is using them). People want to play with Mickey Mouse in new ways, or expose the work of Jane Austen to zombies, or have superheroes that can interact across corporate owners.

The process of modification and change in ancient mythology is almost a sort of fancy fan fiction, but with the possibility of sanction and imprimatur from the broader society. And so, too, does it go with open source software. When it comes to Greek mythology, which of these stories is the true version: Ovid's *Metamorphoses*? Johnston's *Gods and Mortals*? The D'Aulaires' delightful stories and illustrations? *Percy Jackson and the Olympians*? (Okay, maybe not the Percy Jackson version.) There is not a single authoritative version for this kind of storytelling; all of these can be the stories of the living gods.

The same thing is true with open source. Sometimes there is a more "official" version, one blessed by the maintainers of the software. But in other cases, if a "fork"—a version of the original software—becomes popular enough, it grows up and sheds its adolescent awkwardness. Patchy facial hair and the voice cracks of liminality give way to a respectable piece of software. You can see this sort of thing in those many flavors of Unix, from BSD to Linux. Each one is a variant of the original Unix idea, changed and modified over time.

You can also see this in the world of folktales. Many fairy tales, from "Rumpelstiltskin" to "Beauty and the Beast," are not one-of-a-kind stories. They have been told and retold for generations, changing over time. These stories are so common that researchers have created archetypes for them and even used techniques from evolutionary biology to create phylogenetic trees—branching trees of the evolutionary trajectories of these stories—in order to determine the basic features of a tale and how it has changed over hundreds or even thousands of years.

One of the most told tales is "Cinderella," which has been subject to ramifying differences—depending on the story, the lost slipper varies in its material, fairy assistance might not be there, and so forth—since its origins over a thousand years ago. There was even a book published in 1893 titled *Cinderella: Three Hundred and Forty-Five Variants of Cinderella, Catskin, and Cap O'Rushes, Abstracted and Tabulated, with a Discussion of Mediaeval Analogues, and Notes.*

Whether stories or software, there is the ability to take the best of something and work to make it our own. Open source provides the mechanism for recombination as well as fracturing, splitting, and joining over time to suit the needs of each developer. This is the true power of open-endedness.

The video producer Kirby Ferguson made a wonderful series called *Everything Is a Remix*, showing the ways in which our popular culture is a bubbling stew of borrowing and recombination: *Star Wars* is based on a combination of Joseph Campbell, Flash Gordon, Akira Kurosawa, and westerns. Quentin Tarantino's *Kill Bill* is a mixture of a ton of movies. Rap is built on sampling other songs. Nothing is truly entirely original, but when things are combined in novel ways, we can get something new. This is, and always has been, the way of creativity.

THE OPEN-ENDEDNESS OF SOFTWARE

About a decade ago, the book *Why Greatness Cannot Be Planned* by computer scientists Kenneth Stanley and Joel Lehman extrapolated from their work in artificial intelligence. They argued that when an individual, organization, or complex AI system is attempting to accomplish a large and difficult goal, explicitly aiming toward that goal can often be counterproductive. In essence, their argument is that in a complex space of possibilities, aiming directly for some objective leads to cul-de-sacs or unfruitful directions. For example, if you are the inscrutable process of evolution and want to evolve flying birds, you might first want to create feathers, a seemingly unrelated development evolved perhaps for temperature control. But then these feathers get co-opted for the entirely unexpected purpose of flight. So, too, in technology and innovation. Stanley and Lehman examine how vacuum tubes—an essential component of early digital computers—were not invented for this purpose at all. But once in hand, they could be used by computing pioneers to build those large machines.

Stanley and Lehman write of the importance of focusing one's quest on novelty and interestingness. When new capabilities are developed—what they call stepping stones—they can be productively combined in exciting ways and used to traverse the deep waters of innovation, slowly making one's way toward the eventual goal.

This is the power of open-endedness in a tradition of text, from stories to source code: when basic features are available for recombination, unexpected uses will be gained, whether it's ancient gods or open source packages. The key here is having as large a set of stepping stones as possible.

A powerful hallmark of a successful piece of software is when its creators are surprised by what users can build. For example, Hopscotch is programming language software that provides a visual method for writing computer programs and is aimed at middle-school children. The creators of Hopscotch told me that they were constantly surprised by what these kids build, from games to beautiful graphics (a related project, Scratch, has allowed its users to recreate *Super Mario Bros.* within it).

Of course, not all surprises are pleasant. The creators of the programming language BASIC were disappointed in how that language escaped their control, absorbing a whole host of features that they never sanctioned or desired. Their pristine initial vision was overtaken by what they derisively described as "Street BASIC," a kind of "vernacular" that varies "from machine to machine and year to year . . . full of ugly hardware-dependent features that are hard to understand and use." Not what the creators envisioned, but what seems to have won out.

Ironically, some open source projects are tightly controlled in the quest to prevent the creation from being hijacked. For example, Guido van Rossum, the creator of Python, was given the joke title of benevolent dictator for life, though he abdicated in 2018 and there is now a small steering council that controls the development of the language. Linux development is still controlled by Linus Torvalds, and he is the only one who can allow contributions to become part of the main project.

In the end, though, if creating open-endedness is the goal, giving up a certain amount of control is the key. Some degree of mess is part and parcel of open source.

———

And yet, the hurly-burly of open source is not all sunshine and protocols. Just as a mess can cause vulgar forms of Street BASIC, this wild open-endedness and combining lots of different pieces together can be far from an unalloyed good. There is an adage in the world of open source that with enough eyeballs, all bugs are shallow. This is the idea that as more and more people are involved in the process of building and maintaining software, bugs and glitches are found and repaired. It's almost viewed as a physical law: the more developers are involved, the more likely the bugs are to evaporate and for the software to achieve its perfect form. The textual equivalent might be that with enough people looking at a text, all the typos will have been eliminated.

But this is not how open source works in reality. Far from this Platonic ideal, most widely used open source software has a big disparity between the number of users and the number of maintainers. Even massively popular software might only have a handful of people involved in maintaining it, and they are often overworked and under-resourced. As a result, bugs continue to exist. There have been bugs lingering in code essential for the operation of the Internet, but no one finds them for years or even decades. For example, in 2019 a security flaw was fixed in a set of open source tools for connecting one computer to another known as PuTTY. How long had this flaw been in the software? Since 1999. Two decades! In another instance, a twenty-five-year-old bug hung around in one flavor of Unix before being corrected.

Part of the reason for these persistent bugs is that sometimes programmers hope that something will be fixed or reworked, and that the code they write is nothing more than a placeholder for the time being. But then other things happen, and programmers move on, or it's forgotten. And then it just stays there. A bizarre example is that some of the lines of computer code in the Apollo mission software are actually labeled "TEMPORARY, I HOPE HOPE HOPE." And yet it was still used to get us to the moon.

This is also related to the fact that the ideal of maintenance is rarely achieved. When software is interconnected, our ability to understand how it all hangs together is reduced. And this is far from a new problem. In 1976, the computer scientist Joseph Weizenbaum articulated the issues around trying to wrangle large and complex systems: "These gigantic computer systems have usually been put together (one cannot always use the word 'designed') by teams of programmers, whose work is often spread over many years. By the time these systems come into use, most of the original programmers have left or turned their attention to other pursuits. It is precisely when such systems begin to be used that their inner workings can no longer be understood by any single person or a small team of individuals."

This is very much the world that we live in, one of interconnected systems that are far less understood than many of us might realize (and which I explored in my previous book *Overcomplicated*).

As we build software to operate a modern society, we must weave between two extremes: novelty and maintenance. We can learn how to grapple with this balance by looking at how religions approach tradition. For example, Judaism, by combining a deep reverence for history and text—text that can be drawn upon in times of catastrophe and rapid change—with the understanding that each generation needn't be content with just revering the past, has created a distinctive mechanism for *creating* while also *maintaining*.

Elsewhere in the ancient world, the plans of the gods were rarely to be questioned or violated by mortals. When a god was defeated or outsmarted by a human being, very little good could come of it. For example, in ancient Greek mythology, the warrior Diomedes wounded Aphrodite in battle, and she in turn caused his wife to be unfaithful. Sisyphus outwitted the gods numerous times, only to be eternally punished in the afterlife. Divine fate and wisdom were not to be questioned; they were simply to be endured. There is a divide here between the creator and the maintainer: the gods articulate a plan, and humans are merely bit players in its execution.

But ancient Judaism provided a different vision. In a famous

Talmudic passage, God intervenes in a legal dispute, only to be overruled. God's response? "The Holy One, Blessed be He, smiled and said: 'My children have triumphed over Me; My children have triumphed over Me.'" Imagination or pushback in the face of the divine might be foreign to other ancient religions, yet not for Judaism. While Ovid wrote of a mythical creature—the Ophiotaurus—whose entrails, if burned, provided the rather obscure means of defeating a god, Judaism provides a much more straightforward path: just be good at thinking about the Jewish texts and their relationship to the world. Each generation might be tied to a tradition, but it is a tradition that is made one's own.

This is how we must view open source, as a tradition to be added to, questioned, and built anew in each generation. When we have so many stories retold over the generations and texts preserved across the millennia, it is not so difficult to mine an archaeological tell of wisdom to deal with new problems. What does Judaism have to say about lab-grown meat, for example? Plumbing the depths of Jewish literature, we find multiple relevant tales from the Talmud, whether it's the creation of a golem-like animal or meat that has descended from heaven. So, too, as the lore of GitHub—an online repository for developers to publish, share, and remix their code—grows ever larger, there will be more and more that can be drawn upon in creating what we need over time. This is what a good tradition does.

In the Iranian city of Yazd, on a bustling street, sits a Zoroastrian temple. Within this small building is a sacred fire that has been tended to for over 1,500 years, brought to its current location a little less than a hundred years ago. These flames have been maintained for these many centuries. Each generation takes this fire—this oldest of technologies—and tends to it but also changes it. For each fire is ever a newborn.

Alan Perlis—an early computer scientist and who was involved in making it an academic field—was known for his many aphorisms. Among these pithy statements is this: "Every program is a part of some other program and rarely fits."

The creation of software—this profoundly textual tradition—relies on the features of how myths and texts grow and change over time. As we combine and recombine software, we must strive to make it fit, allowing this strange fire of ever-changing text to persist for the long term.

———

But what of the strange fire itself? We've explored how code works and its magical properties, and I've hopefully provided a clearer understanding of how software develops over time. But when all of this comes together and combines with the mind-boggling speed of the digital computer and its mathematical operations, this strange fire needn't remain a small flame. Because of the interaction between code and the machine, complex and beautiful worlds can blaze forth. That is what we explore next.

5

A Universe in Ten Lines of Code

The Realm of Tireless Calculation

An ancient rabbinic text declares that with ten utterances the world was created. Based on the story of Genesis and its steady drumbeat of "And God said . . ." it was determined that there were ten statements that led to the creation of the universe.

One could reasonably question why just one utterance wasn't enough for an omnipotent God. But let's set that aside for now and ask something different: If code provides some of these godlike powers, and we settle for a handful of lines of computer code, could we do the same thing with software?

Or, more humbly, can large realms of complexity unfurl from relatively small snippets of code? It turns out that such realms can indeed arise from computer programs, from *emergent microcosms*, as I like to call them. The computer is the real-world version of the genie's description of himself in the movie *Aladdin*: "Phenomenal cosmic powers, itty-bitty living space." Combine software with a microchip, and you can create the world.

We've already explored the elements of magic in code as well as the textual traditions for how software is developed. But to truly see the implications of all of this—how the computer is something truly different and how software does its thing—means exploring the possibility of emergent microcosms. Because these are only possible due to the digital computer's capability for relentless calculation.

Humans tire easily, but not so machines. I'll get bored and make mistakes if I'm supposed to multiply number after number together. But computers have no such issues. A computer is a device that, as the physicist Richard Feynman noted, "is as dumb as hell but it goes like mad!" Because of this unrelenting and unceasing ability, computers are able to reveal the implications of equations or algorithms beyond the ability of any person. Computers can graph long streams of numbers onto a grid or calculate equations further than any person has ever done. This is the source of the computer's power: it can update every pixel on the screen multiple times a second, it can process huge text files with a click, and it can search the Internet without rest.

Simplicity when unrolled through the work of a computer's tirelessness doesn't just give us lots of simplicity but can actually create something new and wonderful. In particular, computers can elaborate the complex implications for a small snippet of code and unfurl a beautiful cosmos within them. The features of mathematics can be revealed through the untiring operations of a computer—the implications of equations expanded—and, in turn, allow us to visualize these complex worlds. To be clear, these are not simulations; we will get to massive simulations inside computers later in this book. These are elaborations of mathematics through the never-wavering computations of our machines.

For example, in the video game *No Man's Sky*, while the program itself is large, its entire virtual universe is determined by a single 64-bit number. This number acts as the numerical seed that's plugged into the code, which then determines the specifics of the game's sprawling cosmos, replete with stars and planets and alien life. It's an example of the process of procedural generation—combining mathematics and geometry with clever algorithms—that allows for huge complexity to be unspooled from computer programs.

We now turn to the ways that this all happens. But to understand the open-ended creative potential here, we need to go back a few decades in computer history to the screensaver, a genre of software now nearly extinct but that exemplified these emergent microcosms that exist at the intersection between mathematics, computation, and art.

FROM THE SCREENSAVER TO CREATIVE CODING

The screensaver was of the era when computer monitors were large, beige boxes housing massive cathode-ray tubes. These displays had a specific shortcoming: when an image, such as a login window, remained on the monitor for too long, it would "burn in" to the screen and remain there, a ghostly reminder of what it had displayed for long stretches of time.

Screensavers were developed to solve this problem. After a period of inactivity by the user, the screen would turn black or display an ever-changing pattern. Due to the desire that the graphics shouldn't repeat, screensavers required an algorithmic method of generating graphical novelty. Computer programs were developed that imagined one flying through space as stars streamed by, or watching fireworks explode, or observing a horde of winged toasters. These and so many more were made in ways that relied on digital randomness combined with mathematically unspooling more complex graphics. These screensavers melded mathematics and creativity in a way that I—as a young computer user—had never seen before. And I was enthralled.

I loved watching these screensavers, and I collected as many examples as I could, primarily ones created by Berkeley Systems, the producer of the flying toasters. After Dark, the company's screensaver collection, included a screensaver game show, a screensaver video game, and a screensaver fish tank. I even tried my hand at building this kind of program. In high school, for my final project in a computer programming class, I wrote a program that made the screen look like you were flying through a cluster of stars (and in the process, I had my first taste of how useful trigonometry could actually be).

There were many screensavers that were even more overtly mathematically derived. I first saw the word "Lissajous" in the context of a screensaver, only learning years later that it also referred to a specific type of looped wavy pattern, one that can be created on an oscilloscope. There were screensavers that borrowed from realm after realm of mathematics, all in the service of producing images of surpassing beauty. There were even screensavers of fractals, including, of course, the most famous fractal of all: the Mandelbrot set, a series of brightly connected circles and blobs that give way, upon investigation, to an infinity of zoomed-in detail. It's a kind of *Alice in Wonderland* rabbit hole, but built from calculations. The deeper you dive into this digital space, the greater the likelihood that you see an image that no one else—whether mathematician, computer scientist, or simply amateur fractal spelunker—has ever before seen. In fact, Benoit Mandelbrot, he of the eponymous set, has noted that he was only able to do the work he did because of his use of computers (he was employed at IBM). Without computers, that particular complexity would never have been revealed.

Screensavers are no longer necessary—modern LCD screens don't suffer from the same kinds of burn-in issues as cathode-ray tubes—and the screensaver software mania ebbed somewhere by the late 1990s. There are still screensavers, of course, but they no longer have the same hold.

Into this gap has stepped creative coding, a porous realm with fuzzy borders, best defined as computer programs designed to be artworks. Programs are built in the mold of the individual artist and their creative vision, using a combination of mathematics and aesthetics to generate graphics and animations of thoughtful beauty. There is a lot of science here, for sure—an obsession with fractals or emergent behavior is something found deep within the creative coding community—but it is less about research and new knowledge than it is about playful exploration.

In 2001, not long after the time that screensavers were losing their mojo, the programming language Processing was introduced by Ben Fry and Casey Reas. Processing was an attempt to develop a complete programming environment aimed toward visualization for all sorts of nontraditional programmers, from engineers to designers and artists. It was based on Java, the hot programming language of the time.

Writing a computer program that does anything with text has long been straightforward. But if you want to paint the screen with colors, or make an animation, or do anything visual, you have to overcome hurdles such as installing certain packages or interacting with the screen in more complicated ways.

Processing eliminated all of that complexity. It made it straightforward to draw a circle, for example: by simply typing `circle (100, 100, 50)`; you get a circle with its center at a hundred pixels from the left of the window and a hundred pixels from the top, with a diameter of fifty pixels wide.

You could easily make animations using Processing, repainting the screen quickly and cleanly as your graphics changed. Processing was adopted by the artistic community, attracting a huge number of users, with the language still around and thriving twenty years later. I used Processing in my computer science undergraduate senior thesis as part of a project that created an abstract model of the origins of life, with some fun visuals of different-colored beads on rings.

Over time, Processing spawned numerous dialects—one using Python, a version in JavaScript, one even using a variant of Clojure—and created a large community. If the creative coding community has a lingua franca, it would likely be some form of Processing.

What have people made with it? Well, certainly a lot of programs that look like screensavers. There are huge numbers of whimsical graphics created by users that display geometric, biological, and science-fiction-like images. There are programs that have a mandala or meditative quality to them, and others that feel cartoonish. There is simulated snowfall, simulated fire, and even simulated rotating Rubik's Cubes. The designs behind these involve messy mixtures of artistic decisions and mathematical constraints. In many ways, these programs feel like the collective beginnings of articulating an entirely new aesthetic.

The creative coding community, enabled by Processing and similar tools, has re-enchanted programming. These programs are often not particularly long—they can be made by a single person, holding the entire thing in her head—and they have hints of magic when done right, generating beauty and operating in ways not immediately apparent. They allow for a certain amount of creative expression. There is an art and personality to creative coding that is not as clearly visible when developing enterprise software.

And they are able to do what they do because of the combination of mathematics and the unflagging power of computation. Want to generate falling snow? You can easily generate thousands of small white particles, each with its own behavior as it falls to the ground, due to the power of a computer program able to update each one many times per second. Each falling dot on the screen might have a particular location, speed, and maybe even direction. The program is able to specify how each particle changes these values at every moment in time, and the computer happily complies. An individual flake might move down a couple pixels, then across, then down again, something that by itself is unremarkable. But when done over and over, and rapidly and on a large scale, this goes from a wandering single circle to a blizzard. The machine handles all of this, something that would be nearly impossible—or at least unbelievably tedious—to do by hand. Creative coding and screensavers rely on a computer to expand a small description into something beautiful.

These features are also seen in two other somewhat overlapping communities: the demoscene and code golf.

The demoscene (pronounced "demo scene," not "demo-skeen," as I have to continually remind myself) is a subculture devoted to building the most impressive "demo," or short computer program for audiovisual display, within certain limits, such as one that is only 4,096 bytes long, smaller than a modern empty Word document file. Into these constraints have stepped virtuoso displays of entire rendered cities and impressive musical creations. The demoscene is replete with programs that push the limits, showing what can truly be articulated from these small sizes.

A similar kind of work is done in the code golf community, which also values brevity, but for writing computational algorithms. Just as golf involves getting the lowest score possible, those competing in code golf count the individual number of characters required to accomplish some computational task. Economy of description is taken to a new level here. Entire programs might fit within a single line of impenetrable code, little more than a baroque string of symbols. For example, using the programming language PHP—normally used for programming on the Web—you can calculate the mathematical number pi, albeit slowly, using this string of code (in case you're wondering, it uses something called a Monte Carlo simulation to converge on the value of pi):

```
for($i=1,$j=$k=0;$i++){$x=mt_rand(0,1e7)/1e7;$y=mt_
    rand(0,1e7)/1e7;$j+=$x*$x+$y*$y<=1;$k++;if
        (!($i%1e7))echo 4*$j/$k."\n";}
```

While code golf focuses particularly on economy of code, when it comes to generating images and figures, the question becomes, how is all this variety generated, especially if the programs are far too small to store actual images or videos? This leads us to a fundamental idea of creative coding and its harnessing of the tireless computer: procedural generation, something we've already hinted at in terms of the power of mathematics mingled with code. Procedural generation can be best understood by delving into the idea of the fractal.

FRACTALS AND THE COMPRESSION OF COMPLEXITY

Ever look carefully at a tree's shape? Not the preschool version, with a trunk augmented by a few stray branches and surrounded by a green cloud of leaves. That's not what trees really look like. A tree is a recursively branching structure: each branch divides into other branches, which in turn fork into more branches, and so on, until it gives way to leaves. This self-similar structure is a fractal, one of many types of mathematical shapes that were popular in the 1980s and '90s, with their crenellated patterns and psychedelic coloring.

Despite their variety and complexity, many plant structures—because of their recursive mathematical nature—can actually be described by a small snippet of text. Using something known as a Lindenmayer system, or L-system, a set of rules and associated grammar can be constructed to generate whatever type of vegetation you might wish. Want a wispy prehistoric fern? There is a set of rules for that. Want a barren and leafless winter sapling? That can be accommodated too. From wild carrots to the head of a sunflower, all of this biological complexity can be encoded within a mathematical description.

L-systems are not so far from the programming language Logo, where the user employs a series of commands to control a triangular "turtle" on the screen. L-systems operate by specifying a set of symbols, along with rules for rewriting these symbols. Each time the rules are rerun, the string you have lengthens and becomes more complex over time. And what do you do with this string? Well, the string of symbols is a set of instructions for your Logo turtle to follow, tracing out the shape to be drawn.

Let's say you want to make a simple branching tree. You start with a single letter A and a rule stating that A is replaced with L[−A][+A] (L means draw a straight line, − and + mean rotate the turtle different directions, and the brackets are for storing the location). As you keep on rewriting the symbols, each branch is converted into subbranches, which are branching as well. The recursive nature of nature itself can be modeled in code.

This is just one example of what is known as procedural generation: the process of writing a small computational algorithm able to unspool a tree, or an entire city, or even whole planets. The idea is that a small chunk of code, perhaps along with some randomness from a numerical seed, will unfold all of this variety, using a set of rules that these systems hold in common. L-systems combined with some noise can elicit not just a single tree but entire groves, forests, or arboreal biomes. They are able to do so because they have the unfailing patience of a machine, able to calculate over and over, something that humans simply cannot do. As we've already noted, early ideas around fractals and the Mandelbrot set might not have even been developed if it hadn't been for a partnership with computers.

Procedural generation is often used in the computer game industry, where rather than scan an entire world, or even just a forest, and place all this highly complex data into the program, you can computationally describe the near-infinite variety in code and then generate whatever is desired upon command. This is what was used in *No Man's Sky*'s 64-bit number, which provided the basis for its gaming universe. These computer descriptions involve a balance, compressing the fantastic diversity of the real world while still preserving something that looks incredibly rich and realistic. There is a verisimilitude here: things must look real, even if beneath the hood there is nothing but tricks and hacks. While it would be unsurprising if some of the spirit of these L-system rules is embodied in the developmental process of plants, nowhere in a tree's genome is a string like L[-A][+A].

This balancing act of procedural generation results in computational wonders. There are methods for generating maps for the worlds of fantasy stories, which might begin by dividing a two-dimensional space into a series of irregular cells or by using simplified models of erosion. It is possible to generate an entire city crisscrossed with streets and populated with novel and imaginary high-rise buildings. Shading into simulation, computation can even be used to determine where cities might be founded on a map of the United States if one were to rerun the settlement of the country and movement westward by its citizenry.

The fundamental insight of procedural generation is that informational diversity—the richness and variety of some aspect of the world—can be compressed into a more compact description. This parallels an insight of the Scientific Revolution: that mathematical models of reality can unify a wide variety of phenomena, from the law of gravity bringing together the movements of planets and ballistic missiles, to Snell's law, which explains how light moves when passing between different materials.

This is closely related to the concept of "hack value," a pseudo-precise calculation described in *The New Hacker's Dictionary* for graphics as "proportional to the esthetic [*sic*] value of the images times the cleverness of the algorithm divided by the size of the code." In other words, the more compact and clever a description for a complex image or phenomenon, the higher the hack value. And the fewer lines of code required to generate more complex phenomena, the better. Whether it's higher hack value, or a more parsimonious scientific explanation, or a more sophisticated procedural generation within a creative coding program, ultimately these are all in search of the same thing: mathematically terse spells that can unspool massive complexity.

To quote one programmer active in the demoscene on what magic in the world of technology might mean: "It breaks people's preconceptions of what is possible. In order to challenge and ridicule today's technological bloat, we should particularly aim at discoveries that are 'far too simple and random to work but still do.'"

This decompression of simple code into unfurled complexity by virtue of the unceasing computer is something that has been there all along in computing, whether we call it screensavers or creative code or the demoscene or something entirely different. In the 1970s and '80s, magazines and books were full of computer programs—often in BASIC—that generated delightful images and simple games. One of the foundational magazines from this time was actually titled *Creative Computing*. It was published from 1974 to 1985 and provided examples of puzzles, games, and visualizations.

And such a program could even be as elegant as a one- or two-line program:

```
10 PRINT CHR$(205.5+RND(1)); : GOTO 10
```

What does this well-known snippet of code do? All it does is randomly choose to write either a forward slash or a backslash, over and over. This coin flip of a choice, tirelessly executed and then written onto the screen, sounds simple. And it is. But something happens when this randomness is mingled with the computer's ability to perform that which a human would grow weary of nearly instantly: a random collection of forward slashes and backslashes cascades across the screen, which ends up resembling a labyrinthine pattern. This simple program is so amazing that there is an entire book devoted to this little bit of code, full of its variations, explorations of randomness, and other jumping-off points.

But we can see hints of this desire to generate visual complexity from a mixture of randomness and order even earlier. Artists, almost as soon as they were granted access to computers—even with no more than punch cards on a large machine—began to use them to create these unfurled worlds. For example, Bell Labs, that hotbed of Unix creation, was also home to early artistic experimentation with computers. In 1965, a riff on Piet Mondrian's work *Composition with Lines* was developed, using computationally generated pseudo-randomness, and titled *Computer Composition with Lines*.

One fascinating feature of these kinds of computer programs is that they often need to be run to even determine what they will do. There is a surprise for the programmer, where it's impossible to figure out ahead of time exactly what the images might look like without generating them. For example, my friend Max Bittker has built an open-ended sort of drawing tool called Sandspiel. Max did not expect that two of the types of "sand" that the user can draw with—fungus and plant—would together generate a kind of intricate latticework pattern. Only by testing and playing with what he had created did Max make the discovery. Likewise, AI researchers have been surprised by the abilities of their systems when they scale up their models. (They can sometimes do math! And code!) This is similar to a fundamental idea within computer science known as the halting problem: it is mathematically impossible to write a program that can be fed into any computer program and determine whether that piece of computer code will stop running—halt—or if it will run forever. The only way to find out is to run the program itself (and even then, you might have to wait forever). When it comes to computation, you might never be quite certain what you will get. But therein lies more than enough space for wonder and delight.

———

Emergent microcosms begin with computation and the ability to do a lot with a small amount of code. They combine randomness with the cranking out of arithmetic result after arithmetic result, all of which adds up to huge intricacy. There is more than a bit of abstraction here as well: more powerful computer programs are only possible because of the vast foundation upon which they draw. The infinite patience of computers allows a kind of intricate complexity to be revealed.

There are now collections of programs that can fit inside single tweets, capable of unfurling beauty from these small snippets, such as a mind-bending, wormhole-like spiral or hypnotically undulating lines and shapes. Each of these—in a mere burst of text—rapidly constructs an entire world, much as how John Milton's Pandæmonium, the capital city of Hell in *Paradise Lost*, "rose like an exhalation." And in the process, these programs allow us more of that wondering stance within computing.

Steve Jobs spoke of computers as bicycles for the mind (something to which we will return). The idea is that, just as a bicycle allows you to go so much faster and farther than walking, so, too, can computers provide efficiencies for the creative powers of your mind.

Computation itself is a magical bicycle. Through repeated mathematics, small snippets of text can act as world-size levers upon a fulcrum, generating entire emergent microcosms.

PART II

Thought

6

Machine Linguistics

The Beauty of Programming Languages

Biblical prose is distinct. The renowned scholar and translator Robert Alter has argued that the main narrative mode of the Hebrew Bible employs a specific register, a particular style of text. Among the many features of this kind of biblical prose is spareness: it uses a relatively small vocabulary, avoids synonyms, and likely uses words that are more formal than those that were in ancient everyday usage. It also uses a deluge of repetition. As Alter has described it, biblical narration is one of "slight strangeness."

These are all features that Alter has tried to honor in his translations. For example, here is his translation of the first verses of Genesis, describing the first day of creation: "When God began to create heaven and earth, and the earth then was welter and waste and darkness over the deep and God's breath hovering over the waters, God said, 'Let there be light.' And there was light. And God saw the light, that it was good, and God divided the light from the darkness. And God called the light Day, and the darkness He called Night. And it was evening and it was morning, first day."

Whether the precision of the phrase "welter and waste" (an attempt to capture the alliterative *tohu vavohu* in the ancient Hebrew) or the repetition of "and" and "light," this prose is distinct from our own.

As I was thinking about Alter's discussion of the features of biblical narrative, I was initially drawn to making a comparison between biblical prose and computer code. Because, at least when it comes to a relatively small vocabulary—variable names, repeated keywords, specific operators, particular functions in widely used libraries—this is true. For example, as per its official documentation, the language Python has only thirty-five keywords, including "return," "finally," "global," and "False."

But of course, the analogy breaks down very quickly. Human natural languages are not the same as programming languages. While they are both languages of a kind—sort of estranged siblings in some fuzzy linguistic realm—and they both have specific syntaxes, as well as meanings for their words and phrases, natural language tolerates ambiguity in a way that computers do not. There is a messiness to the way humans communicate with each other (for better or worse), that formal machine language is incapable of handling.

There is also the question of purpose. Biblical prose is narrating a story or conveying laws, and a computer program is designed to do something (interact with the user, process some data, display some graphics). Code needs to act. Furthermore, biblical texts were meant to be read aloud and understood within a certain environment, and source code is meant to be looked at in a very different context. Code also has to work in a specific and delimited way. The programmer and novelist Vikram Chandra cautions against making these comparisons, writing in *Geek Sublime*, "To compare code to works of literature may point the programmer toward legibility and elegance, but it says nothing about the ability of code to materialize logic." In other words, code can do more than prose can, such as precisely describe algorithms that operate in the world.

We could continue noting differences. Nevertheless, there is something here, imperfect analogy though it might be. For the way language is intertwined with the mind is complex, something true for any text, from scripture to source code. And that is the subject to which we now turn: the nature of programming languages—these methods of communicating with our computers to make them do our bidding—from the perspective of language. Specifically, just as all human languages are fundamentally able to express the same ideas and thoughts, the same is true in the realm of programming, even though there is a diversity to computer languages. When it comes to the many modern languages that programmers use for writing code, as we will see, they are all fundamentally equivalent.

This all requires recognizing something that is a bit hard to reconcile: programming languages are meant to be read by both humans and machines. These are languages designed for people to write and read, even though they are also fed into computers. They are languages that programmers think in. They are also the raw material for machines to do their thing. As a result of this contradictory nature, computer code is beautiful and messy and powerful, with the upsides of natural language and the downsides of a shopping list.

To understand these features of programming languages, we must approach all of this with a combination of linguistics, history, and literature. For example, there are specific features of computer programs that might be—even a tiny bit—like prose. Just as literary texts have specific properties and rhythms, it's fruitful to compare programming languages based on their literary properties. Is there a rhythm to a Python program or something written in Lisp? And what are the nuances behind keyword choices within each programming language? Despite the limitations on syntax and vocabulary that each language contains, the resulting code has a particular flavor and style.

There are indeed literary differences between programming languages, at least to a degree. Some languages have a pristine beauty but a structure that makes them more difficult when writing practical programs. Others employ syntactic sugar—optional alternate ways of expressing something—with the abandon of an unsupervised child in the kitchen (hello there, Perl). Maybe Python, with its emphasis on white space, is more decorous in its diction and appearance than JavaScript. And the pointers of C programs might have the telegraphic feel of Ernest Hemingway.

The renowned computer scientist Edsger Dijkstra, whom we met earlier, recognized this decades ago. In his acerbic style, he did a vicious takedown of FORTRAN as an "infantile disorder," that BASIC causes programmers to be "mentally mutilated," and that COBOL "cripples the mind" and is a "disease."

Programmers may disagree about my characterizations of different languages—not to mention Dijkstra's—but one thing is clear: each language has an aesthetic, a feel, a vibe. There are cultures that arise within each language, and programmers become emotional about their personal favorites.

Languages of course conjure up these feelings, but judging each language in its entirety is not really fair. Some programs in Python can employ the more functional features of the language, giving it a different flavor, or another program can suffer from being entirely tangled-up spaghetti code. Each program—and each programmer—can have its own style, something distinctly personal, including an inordinate amount of bitching and moaning in the programmer's commentary within the source code. Even within the strictures of programming, a personal style seeps through.

Code can be beautiful, or at least interesting and informative. But it is also intensely personal and replete with style. The code programmers write is human, because it is fundamentally textual and linguistic, with many of the same features of our natural languages. To see why, we need to examine the evolution of code over time.

FROM BINARY TO BABEL

Binary numbers are the Adamic language for our computers. While computers don't need to use binary—the first general-purpose programmable electronic digital computer actually used ten-digit numbers, the same way we are taught in school—binary is as fundamental as it gets and is what is used by the machines around us. Using nothing more than ones and zeros, you could communicate with the machine, whispering to it your needs and desires, from addition to memory storage. Feed in a strip of punched paper, or a series of cards with holes describing these bits, and the machine would do your bidding.

But even more fundamental than that, binary is the language of the logic that is responsible for this power: one and zero, on and off, true and false. As we saw back in Chapter 2, everything within our modern computers—when you delve deep enough—is based on these Boolean logic operators. Binary plus logic powers all the features of a computer. But not just any string of ones and zeros will do for a specific machine. The details of a computer, particularly in the early days of computing, affected how it could be programmed.

The first programmable electronic digital computer was arguably the Electronic Numerical Integrator and Computer (ENIAC). Completed near the end of 1945, the ENIAC was the size of a room. How was it programmed? You controlled its calculations by setting switches and wiring up its components. But each program was a bespoke affair. The program itself was the specific arrangement of cables and switches, something that could take days or even weeks to set up.

This form of programming would not stand. People operating computers needed to store a program more easily, whether it was in the memory of the machine itself or in some more straightforward written medium that you could feed into the machine to tell it what to do. Computer programs began to be encoded as instructions in punch tape or onto punch cards, with each card containing a series of instructions and each instruction being a binary sequence. Intriguingly, this development was derived from a type of programmable loom known as the Jacquard machine, which used a set of paper punched with holes to specify which specific textile pattern should be woven. Computing is most catholic in its inspirations, whether mathematics or tapestries.

In addition, each machine had a set of instructions that it was designed for (how it did arithmetic, how it stored numbers). So when binary was used to program a machine, this involved specifying the operation codes for that type of computer, where each sequence of binary corresponded to a specific instruction, in addition to being able to represent binary numbers themselves.

While encoding a computer program onto a series of punch cards was certainly easier than remembering a configuration of wires, it was not without its problems. If you dropped your shoebox full of punch cards, was there any guarantee that you would remember the specific order of all these instructions? This could be a debilitating problem for a programmer. One workaround was "striping": the programmer would draw a thick diagonal line across the stack of punch cards, so that if they were scrambled, they could be more easily arranged back into the proper order. Your goal went from remembering the logic of the inscrutable binary instructions on your cards to simply figuring out how to reassemble a diagonal line.

But whether programming in machine code or even in a somewhat more human-readable version known as assembly, it was both difficult to understand and specific to the machine you were using. For example, if I wished to calculate pi, I would probably need to know how arithmetic worked in the machine I was working with.

And so, from the initial Adamic language of binary came the dispersion of digital Babel, with higher-level programming languages designed to change this state of affairs and eventually exist more independent of a specific computer. The first widely available programming language was FORTRAN. Released in 1957, FORTRAN (a contraction of "formula translation") was designed by the computer scientist John Backus while working at IBM to have a specific syntax and a certain amount of human readability. Once a program was written, it would then be converted into a format fit for a machine. For reading binary is bad for the eyes and for the soul. It's too hard.

This vital fact must be remembered about programming languages: as many have noted, these languages are designed for people, not for computers. That is the reason there is this process of compilation from programming language down to machine code. Programmers want to write something easy to understand but then convert it into something that a machine can run.

At the same time, though, open-ended English—or natural language more generally—has its issues. It is more powerful in many ways than a programming language, able to evoke the emotional resonance of a sunset, what it's like to lose a friend, or even the implications of traveling back in time and meeting your relatives. But, alas, it's not so great at precisely defining an algorithm, since the richness, the ambiguity, and the synonymous wonder must all be drained away. Otherwise, the machine doesn't know what to do.

Human language without ambiguity seems to be the aspiration of most of these programming languages. Within this goal, there is still a broad realm of languages that can satisfy these needs and penumbras of intention that give code its flavor. Programmers can write loops in different ways, use variables in their own personal style, or comment on an algorithm in a particularly entertaining manner. But each language is designed to satisfy certain requirements and desires, and to avoid the mistakes of what has come before it. For there is no stronger creative force than a programmer unhappy with the status quo in programming, filled with a righteous need to create a new programming language.

After FORTRAN burst forth a myriad of programming languages: C, C++, COBOL, Java, ALGOL, BASIC, Ada, APL, Caml, Lisp, Ruby, JavaScript, Python, Perl, Pascal, Scheme, and many, many more. We can even provide a family tree for many of these, examining the begats of each language—C begat C++, and C++ begat Java—as well as the multifarious inspirations for each new dialect and tongue.

But beneath this computational menagerie there is an order. You can see this in how people learn programming languages.

For example, how many programming languages have I used? My memory is imperfect, but I'm pretty sure I've worked with at least the following: Scheme, C++, Java, Python, Perl, BASIC, Pascal, R, PHP, Logo, HyperTalk, ActionScript, MATLAB, Processing, and variants of JavaScript and Clojure. While this litany of languages might seem daunting, let me be clear: I am unlikely to remember the details of nearly all of these and often need to look up how to do certain things. But that's okay, because just as natural languages have similar styles or families—whether it's grammar or etymological roots—so, too, do programming languages, easing the process of learning each individual language.

Most beginning programmers learn a type—or paradigm—of language known as imperative programming, basically a list of commands that changes the current set of variables and the state of the machine. Set the variable x to 5. Open a file. Add two numbers. This is the most straightforward sort of programming language—C or JavaScript, for example—and its style is widespread.

But among the many styles there is another fairly common category of programming language, one that is sometimes taught early in one's formal computer science education: functional languages. Rather than consisting of a set of commands to the machine, the program in functional languages is a set of mathematical and computational functions operating on each other. Imagine you have a list of numbers and wish to square each number. To do so, you simply apply the squaring function to every element of the list, and you're done. Want to add up these numbers? Then you can apply a function that sums every element of a list and returns that number. It's functions upon functions, operating on lists.

These functional languages are elegant to write but can easily devolve into massively nested parenthetical statements. In addition, they make generous use of a process known as recursion, or self-reference, with functions that refer to themselves. Copious recursion has the unfortunate side effect of making the code more difficult to grasp for a programmer who is new to this kind of language.

The classic example of a language of this form is Lisp. It's often viewed as a kind of Platonic ideal of a programming language. I've even seen Lisp described as "God's own programming language," full of sizzling power and unlimited possibility. In fact, the definition and description of Lisp can be written in Lisp itself in only a handful of lines. This description has been described as the Maxwell's equations of software, unfurling something broad and powerful in a compact description. The physicist James Clerk Maxwell described the properties of electromagnetism in four equations, and the computer scientist John McCarthy and colleagues described Lisp and its entire operation in less than a single sheet of paper. But despite its power and elegance, Lisp is simple enough that a language related to it—Scheme—was used in one of my introductory computer science courses.

However, it can be difficult for a mind to grasp this kind of language initially, particularly if you're only familiar with imperative languages. I remember when I had to make the jump from imperative to functional languages my first year of college. There was a period of metaphorically banging my head against the wall to understand this intricate and soaring structure—holding the rising tower of functions in my head—until it finally clicked.

In fact, this is generally the process of programming more broadly. When learning to program, there is an initial process of trying to understand the syntax and the mental model necessary for it. Nothing works, and errors are thrown with abandon. But then the mind begins to grasp the language's idiosyncrasies and logic. And suddenly, you're programming.

I saw how far I had come when a friend of mine was taking an introductory computer science course in college and asked for my help. I was able to recognize instantly that his hamstrung program was faulty because of a missing semicolon. However, this was only possible for me to do because of a long period of practicing and internalizing a mental model of both this language and programming in general.

Many programming environments try to minimize this—helping you see the kinds of inputs a function might take or even providing auto-complete, sometimes augmented by artificial intelligence—but a programmer must still understand a language intuitively and hold the nature of the program in her head, at least to a certain degree. Programming languages end up becoming intertwined with the programmer's own mind.

There are many other modes of programming language as well. There are languages that use objects and message passing, such as Smalltalk or Squeak; those that use something called a stack, such as Forth; and those that are designed for a specific use case, known as domain-specific languages. There is Perl, described by its creator as drawing magpie-like from many other languages: "I lovingly reused features from many languages."

While the similarities of languages are worth delighting in, the detailed differences are also wonderful. It's fun to look at the weirdness of natural languages: Why do we drive on a parkway and park on a driveway? What is the deal with the pronunciations of "through" and "tough"? You see a lot of this in programming languages as well. Testing if two things are equal is generally done via ==, but in JavaScript you add an extra equal sign (===). Many languages employ syntactic sugar, to allow the programmer to write something in multiple ways and often more compactly (as noted earlier, Perl is so replete with syntactic sugar, it is basically Pixy Stix). And while there is certainly a precision to programming, it is still far more context dependent than we might realize, with certain functions of the same name doing different things depending on the situation.

This broad diversity in the realm of programming languages is never-ending, and it is always being added to. Just as the diversity and features of natural languages betray their origins—the grammar of English demonstrates its Germanic roots, and its vocabulary displays Romance-language influences—we can detect the complex process of bricolage in programming language creation.

Researchers have studied the evolution of programming languages, how the begats of computing history yielded this tree of language. But of course, languages—both natural and computational—needn't have only a single influence or source. Objective-C is influenced by both Smalltalk and C. C++ is not just based on C but also influenced by Simula. Each language recombines features of multiple languages, making the traditional form of evolution the improper model. This is much more a case of what is known in biology as horizontal gene transfer, where bits and pieces are shifted from one organism to another, sometimes even across species. An organism can have a complicated parentage.

To be clear, this evolutionary process is very different from biology or even how a natural language develops. In our spoken languages, grammar shifts, meanings of words change, and even pronunciations are altered in an organic and unplanned way. Programming languages, on the other hand, need to be specified explicitly, otherwise the compiler or interpreter will break. It's fine if someone doesn't fully understand '80s slang—"This variable is *bitchin*'"—but if I innovate in a program itself, everything comes to a screeching halt. Evolution is a much more planned and directed process when it comes to programming languages. It is artificial selection, like the breeding process we see in domesticated animals or in agriculture.

Despite this digital linguistic zoo, this fantastic variety of syntax and style, all languages eventually convert to the binary of machine code at some point. This multi-decade efflorescence still results in the bits and bytes of the computer. Due to this, we can discover something profound about computing: its universality.

THE UNIVERSAL
NATURE OF PROGRAMMING

A handheld calculator can do a number of things very well: add, subtract, calculate square roots, and whatever else its buttons promise. But, aside from the sophisticated scientific programmable calculators you might have used for AP Calculus, they are not truly general-purpose computers. Why not? Well, they can't be programmed. You can execute a variety of functions, but you can't give the calculator a set of instructions that it then carries out.

On the other hand, due to the specific architecture of a CPU—the chip lurking inside every computer—true computers have an open-endedness. They can store data that can then be operated upon. And as long as certain criteria are met, no matter how slow or weird one is, all computing machines are in the end equivalent.

This is the property known as Turing completeness, after one of the grand mages of early computing, Alan Turing. Turing described as a thought experiment the idea of an abstract computer, which we now call a Turing machine. It consists of a tape with information on it, and a little machine that can read the tape. Depending on what is on the tape, the little machine moves back and forth along the tape and writes on it.

In case it is not abundantly clear, this is not what the inside of your laptop looks like. Nevertheless, despite being slower and more one-dimensional and requiring eons to run simple programs like the ones we use on a daily basis—and the infinite length of its tape aside—a Turing machine is effectively equivalent to a modern computer. All machines that are Turing complete are theoretically identical. Converting the behavior from one system to the other isn't easy, but it's possible (we saw hints of this already with Tinkertoy computers and machines based on water in tubes and pipes). This is a profound and nonobvious idea: that after a certain level of complexity, all machines that can compute are all fundamentally equivalent. *Minecraft* is Turing complete, but it's not recommended to use it to run spell-check. We saw before that how crabs interact can be used to mimic the logic within a computer. It might work, but this is not the best way to do math problems.

This same kind of thing is true of programming languages. Once a certain set of capabilities are included within a language, anything you can program in one can be done in another, from calculating factorials to searching a database. Lower levels of programming sophistication—as categorized under the Chomsky hierarchy, named after the linguist Noam Chomsky for his work on languages—are limited, but beyond these simpler levels, basically every language is theoretically equivalent.

This is the same sort of thing we see in natural language. At some fundamental level, every language can be expressed in another language. The basic nature of our cognitive machinery allows for language, no matter which one you learn.

Of course, each language is distinct in its own way, from vocabulary to grammar. And we can see a lot of these shades of difference in the art of translation. Due to the vagaries of culture and history, translation is not a one-to-one-word affair. There are idiomatic expressions, the difficulty of preserving style, and even wordplay that's hard to capture. (We saw some of this in the magic conjured from translation in R. F. Kuang's novel *Babel*, where the difference in meanings between languages could generate a powerful sorcery.) The further we are from straightforward statements, the more difficult the nuances are to capture. But there is no situation where thoughts expressed in one language are fundamentally unable to be conveyed in another.

So, too, with computers. Just because methods of programming are theoretically interchangeable does not mean that they are all equally straightforward. Or easy. There is an entire subculture of programming languages known as esoteric languages, or esolangs, that show the truly wide scope of what programming can be. For example, there is a programming language that consists entirely of white space (such as new lines and spaces and tabs), and there is even a language known as Brainfuck, which consists entirely of the characters ><+-.,[]. These are programming languages as provocation, or art, or simply byzantine frustration. Many of these are examples of something known as a Turing tar-pit: a language that is technically Turing complete but far from easy or efficient for building anything useful. Don't fall into these.

Even though all these languages—from Brainfuck to Python—are technically equivalent, there are profound differences. And that is why programmers keep on inventing new ways of programming computers. Just as there is no end to the making of books, there is no end to the making of programming languages. Each language is an attempt to address the limits of the others and to serve an ultimate need: converting the vague and fuzzy ideas in our heads into instructions for a machine. The Apple IIe owner's manual made these differences explicit to its readers, and with far less crankiness than Dijkstra, referring to Pascal as "Structured Sophistication," Logo as "Child's Play," and FORTRAN as "An Old Hand at Science."

The more expressive a language is—the better and more easily it can convey our ideas to the machine—the more useful the language might be (and the more productive the programmer). This is also intertwined with how much mental energy must be expended by the programmer at any point to understand what is going on, or to otherwise hold the logic of the program in her head. Reduce this energy and increase the expressiveness, and you have built a language that sparks with possibilities. Such is the holy grail of languages. And this is why different programming languages conjure up so many emotions.

Attempts to achieve this holy grail have taken many paths and draw from many different domains. There are so many ways to approach this, whether using wisdom gleaned from psychology research to understand how people learn and think or conducting surveys to see how users actually engage with a programming language.

One way to bridge the chasm between natural language and programming is by creating documents that intersperse comments and explanations with the code itself. Known as literate programming, this is an attempt to show how language is part and parcel of understanding code. Another approach is to work to create a visual representation of much of the program's workings—making it into an interactive medium with feedback. This is based on the idea that programmers should not hold so much in their minds, understanding how all the pieces build upon each other, or even the current state of variables. Instead, these intermediate bits should be displayed to the developer as the program operates.

Taking visualization of programming seriously can be done in a multitude of ways. Some strap the text of code onto a live view of what is happening in real time in the program, allowing the coder to see the flow of information and the state of data. Others turn programming into a visual act, connecting shapes together to create a visible logic for what the program should do. Some languages are even designed such that their results are artistic, where the code itself is fit for displaying in a gallery.

There are tiny programming languages and environments where you can write simple functions and instantly see how they create graphics, such as a pattern of dots on a grid, which is constantly being looped over. These languages allow anyone using them to see the relationship between the code and the imagery.

There were even several years when a physical space was publicly available as a programming environment. In downtown Oakland, the computer scientist Bret Victor led a small team to build Dynamicland, where the entire room was a place where programs were written and executed. This team developed a programming language that involved text on pieces of paper, which were identified by cameras and "run" by the room itself. The idea behind this—among many ideas—was that physical interaction with code, printed out on paper, would make the act of programming far more tangible than elaborating text on a screen. Through a series of projectors and cameras, you could place a program onto a table, and it could just start running.

Make a small program that draws falling circles, and then place a purple magic marker into a region drawn above the program's text, and suddenly all of those circles are raining down in purple. Or orange, or whatever you want. There is a deep physicality here to this kind of programming, combining objects, the data being used, the sheets of code that are moved around, and even the movements of the users.

Modern generative AI has also come along, with its ability to elide the already leaky boundary between natural languages and programming languages. And so, over the past couple years, programming has changed yet again. It has become augmented by AI, with sophisticated auto-complete, filling in your code as you write or even generating entire programs based on natural language.

———

The ancient Greek playwright Aristophanes wrote that "high thoughts must have high language." And this has always been the goal of programming languages: to construct a system of notation that will be most effective at converting our ideas into action, balancing expressiveness with precision. But will we ever find this perfect, mythical notation for programs, this language of the electric birds?

This is what C and Perl try to do, and what visual programming has tried to accomplish. This was the initial impulse of BASIC, to ease the path of creating software for everyone. FORTRAN's creation was even viewed as having massively reduced the debugging necessary when writing code. All of these are part of the quest for the notation that will aid our control of machines. There has been much work on how notation can enhance thinking, and you can see this clearly in everything from mathematics (fractions and base-ten numerals) to chemistry (chemical equations and molecular formulas).

There are even hints here of something known in linguistics as the Sapir-Whorf hypothesis, the theory that the structure of different languages—vocabulary, grammar, and so forth—affect, at least somewhat, how we think about and process the world.

For example, I went to see *Superman Returns* with a fellow student when I was in graduate school. Afterward, I asked my friend what he thought of it. Since my friend had grown up in Vietnam, I wasn't sure how familiar he was with the entire Superman mythos: Kal-El, Jor-El, the flight from Superman's home planet, the Kents, Clark's love of Lois Lane, and so forth. My friend liked it, but he was confused by the "blue rock." Blue rock? I asked. After some initial confusion, I realized he was talking about kryptonite. It turns out that the Vietnamese language uses one word for both blue and green, and for borderline shades, my friend would sometimes get confused as to which word to use.

To be clear, using a specific language neither allows you to think certain thoughts nor prevents you from thinking them; our brains are far too general to be rewired by a specific language. At the same time, even if the strong Sapir-Whorf hypothesis is not true, language does obviously influence how you think.

The words that politicians use can pack a profound emotional impact, depending on what they say ("estate tax" versus "death tax"). Languages with different words for colors mean that their speakers will have slightly different abilities in discriminating specific shades and hues, and if a language uses compass directions (north, south), instead of relative ones (left, right), a speaker will know his physical direction more easily. "Dried plums" sell better than "prunes," because of certain connotations, even though these are clearly the same thing. When we use different vocabularies or languages, we are not processing reality in wildly different ways, but language shades the world we experience.

In the realm of programming, there are researchers who are working to answer how the structure of a specific programming language affects how programmers think about the world and what they are able to do in the realm of software. This is Sapir-Whorf, but for computer programming.

Natural language has been achieving the goal of conveying information to the listener or reader for millennia. However, until the arrival of large language models, natural language was thought to be far too fuzzy for computation (and it might still be). Perhaps AI alongside a kind of "pseudocode" is what we have been looking for all along.

Pseudocode is a misbegotten monster that is halfway between English and code, basically computer code with the specific quirks of the machine and the language elided, such as might be found in a textbook when used to describe an algorithm. You can't actually run it, but you can imagine how it works. A programmer doesn't have to worry about the details of printing to the screen or reading from or writing to a file. She just says she wants these things done. And to be honest, this is probably where AI-enabled computer code is going to shine. A programmer doesn't have to spend hours scouring the Web for the specific reason why some function isn't quite interacting well with a program. Programmers can ask for code that can take text and make it all lowercase as well as remove all punctuation, and a large language model will do what they want. This is impressive programming, but it still leaves the programmer the task of figuring out what exactly she wants the computer to do, even developing the algorithm she might require.

Computer code—intended for humans and machines—is both soaring beauty and ugly details. It rivals the expressiveness of English and the precision of a card catalog. Its crackling energy can be coerced not just to persuade or entertain but to act in the world. Just like natural language, programming's diversity—all of its weirdness and details, elegance and historical trajectory—is bound up in a profound universality.

These are the languages of the demiurges of modernity: our computer programmers. They create the computational systems whose tendrils increasingly extend to every aspect of our lives. But programmers should not be the only keepers of these powers, as we will see in the next chapter.

Prometheus must bring computation to all humanity.

7

In the Beginning Was the Spreadsheet

Democratizing Code

When my daughter was in kindergarten, I built a simple piece of software so she could practice her sight words—the common words you have to know cold without sounding out—as she began to learn how to read. Instead of going through a set of small cards on a key ring, she could just tap the keyboard to see the words. It was nothing fancy, just a program that showed words for her to memorize on the screen. But my daughter was delighted by it, pressing the space bar over and over to get the next word.

I'm not an expert programmer by any means, and while this task wasn't hard, it required some work and effort, not to mention the accreted experience of about twenty years of coding. But most people can't do this kind of thing: there haven't really been tools widely available for making lots of types of software without knowing sophisticated computer programming.

But code should not be the exclusive domain of the wizard steeped in lore, the hierophant surrounded by his esoteric mysteries. It is not so distant or baffling, available only to those who can reach the digital heavens. Most everyone should be able to create small tools and applications. These little programs won't be showered with venture funding or be the basis for the next Facebook. But they could help us do our jobs better or make our lives easier or more delightful. Imagine being able to build the simple note-taking app you've always wanted or the drawing program that will make your child smile. However, because this kind of software is hard for nonprogrammers to create, we find ourselves forced to dismiss these desires as not available to us. But it needn't be this way.

Software development is not something just for start-up founders or those working in Big Tech; we can and should be able to quickly and easily create little bespoke computer programs. The part of our mind that can survey the world around us and determine possible software solutions should be open to us all, not just to the programming elite.

This is not an exhortation of "Learn to code." It is a call for making programming easier. We have just explored the languages of computational thinking. And while computational thinking is indeed important, it should be available to each of us. This is a necessary thread within computing: the democratization of programming itself. The history of this expansion of thinking with computers to everyone, even those not comfortable with traditional programming languages, is the focus of this chapter. This is the story of how the barriers to thinking about software creatively can be reduced. And how they already have been, at least to a certain degree.

In fact, I can almost guarantee you've done some programming already. To better understand the power of coding for all of us, we must start in a somewhat unexpected place: the world of the spreadsheet. For if you have used a spreadsheet for anything more than list making, you've programmed, even if in a rudimentary way.

THE SPREADSHEET MINDSET

The beginning of the personal computer truly kickstarted in 1977, the annus mirabilis when the so-called trinity of microcomputers—the Apple II, Radio Shack TRS-80, and Commodore PET—were released. While it then seemed like these machines would become fixtures of homes and desktops, it actually took several years for this vision to materialize.

Instead, there was a curious moment early in the history of the personal computer. While personal computers were popular—sold widely and entering the public's consciousness—they were seen less as useful tools and more as toys, playthings for the technically minded.

All at once, that changed with the release of VisiCalc, a spreadsheet computer program. It was the personal computer's first killer app, the one that people simply needed to have. The spreadsheet gave people an understanding of why personal computers were necessary, and it gave users a reason to buy these machines.

In the years since, spreadsheets have become so ingrained in the world of business that they are taken for granted or simply viewed as documents of annoyance. But spreadsheets are constructions of genius. They are premised upon a single idea, conceived decades ago: they are graph paper that can be executed. Put values in their little cells, and they don't remain inert. A spreadsheet is a piece of paper that actually does something. The functions that lie within make it instantly updatable; it is paper that runs.

Specifically, spreadsheets are a digital version of something from the physical world. It has been forgotten in the distant mists of accounting, but there were once physical spreadsheets—large sheets of paper divided into grids—that were useful for bookkeepers, accountants, or those involved in business forecasting. These spreadsheets would sprawl across a desk, or on a wall, or on the floor. Their massive grids would be filled out, then updated and modified. Every step of the calculation was laid out. This was hard work, but it was also a way of thinking visually.

Dan Bricklin and Bob Frankston, the co-creators of VisiCalc, had been friends for years by the time that Bricklin went to Harvard Business School. While there, Bricklin realized something that was only possible because of both his deep knowledge of computing and his training in accounting: there was a need for creating a digital spreadsheet. These laborious calculations could be transferred onto a computer screen, swapping the manual updating for the ability to have any change in a single cell percolate outward to all other relevant numbers. First released in 1979, VisiCalc made calculation visible (hence the name). With its labeled grid of cells, each of which could be filled with numbers and formulas, it was a basic version of what we still use today.

Originally referred to as an "electronic scratch sheet" or "electronic sheet of paper," and even for a while as an "electronic spreadsheet," the spreadsheet has achieved such ubiquity and such widespread use that it no longer needs a modifier. It has won the terminology war. If you want to talk about the original spreadsheet, you're going to have to say "paper spreadsheet." Computation has dematerialized this type of document. And as a result, it is everywhere.

Mocking up a budget? Spreadsheet. Building a dashboard for how people use your website? Spreadsheet. Figuring out how an investment might perform? Spreadsheet. Spreadsheets have even seeped into the rest of our lives, from being used to plan wedding invitations to keeping track of who's bringing what for a Thanksgiving dinner.

Due to a combination of spreadsheets' ubiquity and power and the fact that they are associated with middle management and corporate speech that borders on the vacuous, when people hear the word "spreadsheet"—or even just "Excel"—there might be a visceral retching that occurs. Of course, this is not always true. Some people love a good pivot table. I am not one of them, though my wife is. She once had an involved conversation with a friend about their favorite respective Excel functions (hers was VLOOKUP).

Because of spreadsheets' omnipresence—along with the fact that what we use them for might not be the most sophisticated situations—we have failed to realize something as a society: we are all programming computers.

When you write a small function that adds a set of numbers, or compares two different columns and graphs the result, or incorporates a formula that averages numbers, this is all programming of sorts. Once users go beyond tweaking the numbers in a sheet to constructing functions that connect the cells in complex ways, whether they realize it or not, they are engaging in a kind of coding.

Computer scientists have thought for decades about this idea of democratizing programming. They want to allow anyone to program a computer themselves—even if in a simple way— and reduce the barrier between imagination and program. There shouldn't be a hallowed elite able to program the machines, with the rest of us relegated to simply using whatever we receive. Writing software is a role for all of us, and we need better ways to make this possible. This urge is seen in the development of the programming language BASIC: the acronym stands for Beginner's All-Purpose Symbolic Instruction Code.

Spreadsheets, though, were the first—and still probably only— widespread example of this democratization in action (this is also referred to as end-user programming, since it allows the end users themselves to program the system). Spreadsheets work particularly well as democratizing forces for computation because the calculations are laid bare. Yes, you often have to click to peer into an equation, but the arithmetic intermediaries can be visible in other cells or other bits of the spreadsheet. At the risk of making too much of it—and also becoming a bit morbid—a spreadsheet is a sort of vivisection of a computer program. You see all of its parts, and, unlike the physical spreadsheet, it's not inert. It can change alongside your assumptions and questions. The magazine *Creative Computing* recognized all of this in its initial review of VisiCalc, instantly appreciating that a spreadsheet was a mechanism for allowing people to program easily.

Spreadsheets were the harbinger of the democratization of programming—something that we need more of—but we can see it happening from time to time in the history of computing, from the early days of the Macintosh to the current day.

THE MAGIC OF HYPERCARD

Even though the Macintosh was released in 1984, in my own personal retelling of this history, its potential was not truly achieved until 1987, when the release of HyperCard confirmed its earth-shattering nature. Much like personal computers were transformed by VisiCalc, this one-two punch of Macintosh and Hyper-Card changed how I thought about computers. It was like nothing else I had seen in computing. The mere mention of HyperCard elicits from those who know it a sense of delight.

Bill Atkinson, its developer, described HyperCard as "an erector set for building applications." Simply put, you could build your own software using HyperCard, with each program made up of "stacks" of "cards." Each card—or screen—could contain text and images, as well as interactive buttons, with the ability to connect to other cards. Think of these stacks as rudimentary websites that existed entirely on a single machine, with each card as a page.

What could you do with these basic features? Pretty much anything you wanted. You could start small, storing and linking information, and slowly build from there. If you were an average user—a nonprogrammer—there was little barrier to building a piece of interactive software easily. You could effortlessly add buttons, text, and images through menus and interactive graphical tools and even provide a bit of code—courtesy of its friendly and readable HyperTalk programming language—to make these pieces all work together. Based on these components, you could make something as simple as an on-screen button that, when pressed, would make the computer beep a few times.

But you could do a lot more than that. You could manage an inventory system or even an entire company. You could build an interactive story, where each page is a separate card and the pieces of scenery are interactive and clickable. You could make educational software, with a stack full of interactive cards on information about outer space or *Moby Dick* or dinosaurs. You could build blockbuster computer games, like *Myst*, which was originally developed using HyperCard. Apparently, you could control the lights of a massive skyscraper: two of the tallest buildings in the world—the Petronas Towers in Kuala Lumpur—had parts of their lighting system controlled by HyperCard. HyperCard was even an inspiration for the World Wide Web, as well as for one of the early web browsers.

Atkinson once described HyperCard as "an attempt to bridge the gap between the priesthood of programmers and the Macintosh mouse clickers." But even more than that, HyperCard didn't compromise between the easily usable and the creatively powerful. All of that was to be found within it. To use ideas often attributed to the computer scientist Seymour Papert—one of the creators of the programming language Logo—HyperCard embodied the concept of "low floors" and "high ceilings": technologies that are easy to begin working with but still have lots of open-ended potential. It provided space for both the beginner and the expert.

HyperCard was a gateway to programming and what first got me comfortable with the idea of coding. It's probably not ridiculously hyperbolic to say that it influenced a large number of future software developers to think computationally. The developer of the original "wiki" software—the foundation for Wikipedia—was inspired by HyperCard. At least one of the modern crop of Apple engineers also credits it for getting him into programming (though I'm sure there are many more). And Samantha John, the co-creator of the children's programming tool Hopscotch, says it inspired the software she helped build.

Simply put, HyperCard was the fulfillment of the truly generative and creative power of the early Macintosh. But there is still so much more that can be done, something that becomes visible by studying Dungeons & Dragons.

THE ARTIFICERS OF PROGRAMMING

In the game Dungeons & Dragons, the kinds of characters you can play are a combination of a large number of types and facets. There are different species, different occupations, different moral alignments, all available for mixing and matching. Hidden among this menagerie of elfin healers, human wizards, and *aarakocra* fighters is a specific category of player who can wield magic known as an artificer. These artificers use their arcane skills to build artifacts: tools and objects that have magic bound up within them but that anyone can use. A staff with a specific kind of explosion, a pistol that can hunt demons, whatever. These artificers take the esoteric and bring it to everyone in a village.

In a talk I discovered, artificers were described within the context of software as the developers who build user interfaces, helping to make things far more user-friendly and bringing certain powers to everyone. In other words, artificers are the ones working on democratizing computation. Spreadsheets make coding simple, allowing some people to build applications on top of them, jury-rigging them to play a game or to be used as an artistic canvas. Artificers built spreadsheets and HyperCard in order to make creating software straightforward and easy.

But artificers have made advances since the 1980s and HyperCard. There are now tools that combine spreadsheets with other kinds of user interfaces and tools that take a kind of visual construction—along the lines of HyperCard—to great lengths. You can now easily build websites or smartphone applications and glue different web services together. There are even tools specifically aimed at this kind of open-ended creation for children. These "no code" or "low code" tools are designed to bring the power of code to the masses.

All of these computational tools are what might also be described as magic crayons. A term from the computer scientist and designer Chaim Gingold, it is a clear allusion to the titular writing implement in *Harold and the Purple Crayon*. Harold uses this crayon to scrawl on the sky itself, creating a boat, animals, an entire city, and nine different kinds of pie. It is a magical device for easily creating objects, allowing sketches to come alive. This is what magic crayons can be in computing as well: mechanisms for creating software, sticking bits and pieces of code together that can then start acting.

This involves finding the right interface and way of structuring the user's thinking, one that allows for the manipulation of software components or data, using the right balance of text and graphics. While user interfaces might be only part of lowering this barrier to computation, an appropriate means for interacting with systems is vital. But what should these interfaces look like? There are those who disparage the graphical user interfaces we are surrounded by, whether it's the desktop metaphor or the buttons on our screens that have multiplied like rabbits in spring. For example, at one point Microsoft Word could display so many toolbars that the on-screen written text would be almost an afterthought. This is not a good thing. But a user interface done well allows you to intuitively understand the possibilities, what can be done and how. The command line interface, for all its power, is a void into which you are thrown. Unless you know the right spells, you are unable to accomplish anything (or, more rarely, you might accidentally delete your entire computer. But let's not dwell on these matters). Would not a graphical interface make it easier to set the time on a simple digital watch? I don't want to fiddle with only three buttons, hoping I understand each one's changing meaning. When created artfully, an interface is not a mask preventing the user from seeing the machine's true nature but an on-ramp for the beginner. It makes doing what experts do that much easier. I'm still partial to having peeks under the hood (command lines, with complete access to the power of code, available when you might need them but leaving them hidden the rest of the time), but overall, expertise requires approachability.

Steve Jobs spoke about how he was influenced by Heathkits, electronics kits designed to be assembled from scratch. You follow the instructions and solder the pieces together, and at the end of the process, you have built yourself a radio or an oscilloscope. In Jobs's words, "It gave one an understanding of what was inside a finished product and how it worked, because it would include a theory of operation. But maybe even more importantly, it gave one the sense that one could build the things that one saw around oneself in the universe. These things were not mysteries anymore." There is a continued need for this kind of approachability.

We need our computational artificers and the magic crayons that they can place into each of our hands. But magic crayons can and should be able to do more: we should be able to "sketch" what we want from our software, describing it at a high level, and then make it so. This program synthesis has been a goal for decades in certain branches of computer science. But this dream of program synthesis has long been a reality in our science fiction.

STAR TREK AND COMPUTERS

Star Trek: The Next Generation premiered when I was in kindergarten. I grew up watching the show, armed with a *Star Trek* lunch box and action figures. The show's details and the nature of its universe are so finely ingrained into my mind that I now find it hard to distinguish between its objective quality and my nostalgia and love for the *Star Trek* world and its characters. There are many features I can go on and on about: the post-scarcity society it conjures, its complicated thoughts on artificial intelligence, the geopolitics of its galaxy. I even first became interested in the ideas of genetics through a discussion my father and I had about one episode.

While I really like *Star Trek*, as I began to learn how to code, its approach to programming always felt a bit . . . off. The crew of the *Enterprise* apparently knew how to program computers, but this training seemed to involve a kind of hand-waving expertise. They often just chatted with their machines. You would ask a computer to examine some data in a specific way, then narrow down the results. Or you could create an immersive experience on the Holodeck by simply describing a location and time period (late Victorian London or 1940s San Francisco), and then you hopped in.

I didn't realize it at the time, but *Star Trek*'s Federation was awash in program synthesis, the ability to generate computer code from a specification. In other words, it was machines generating code for the user. Describe in some way a desired function, and have the computer write it for you. Or give some examples of a small formula or a database query, and have the machine generalize. Or build a data analysis program from a query on the bridge of the *Enterprise*.

For a long time, program synthesis was a firmly academic affair. Understandably so, as the techniques used to make this work had to find ways of spinning the imperfect and messy requests of the human user into the gold of formal and precise computer code. Program synthesis traditionally operated by searching the unimaginably large space of possible programs to find one that fit the specification that it was given. These techniques involved theoretical understanding of programming languages and were often useful for only a small subset of situations, making the dream of speaking to your machine difficult and distant. There were instances where these techniques did break through, with the most celebrated example actually in a spreadsheet. Microsoft Excel incorporated elements of program synthesis with the functionality known as Flash Fill, which can automatically infer the kinds of content that need to be entered into spreadsheet cells.

But AI natural language tools have made program synthesis a possibility, turning it into a supercharged auto-complete—and generator—for computer code.

Imagine you wanted to create an iPhone app with an arrow that, no matter where you were, would show you the direction toward the center of our Milky Way galaxy. Twenty-six thousand light-years away, the center of our galaxy is far less easy to picture than the location of the sun or the phases of the moon.

But what if you don't even know how to make iPhone apps?

This was the situation that my friend Matt Webb was in. Matt is a designer and solid programmer but didn't have any experience with using Swift, the programming language for making these kinds of applications. But he didn't need to, because he could use AI.

There are many options available here. You could use a chat interface, describing what you want and working with the AI as a conversational partner. Or you could use an auto-complete tool for programming as you write code, the tool suggesting lines. Or, when you describe a function, the AI could write entire chunks of code itself.

Whatever you choose, each tool helps overcome the tyranny of the blank screen, the cursor blinking at you in its smug, judging way. Matt opted to use ChatGPT, as it was easy and straight-forward.

He began a sort of conversation with the AI, asking it questions along these lines: "I'm building an iPhone app using SwiftUI. I have installed Xcode version 14. Please walk me through creating a new iOS app with a single screen. The screen should be blank except for a line of text in the middle that says 'Hello, World!'"

ChatGPT responded, unspooling a combination of code and explanation for him. Then Matt continued the conversation, asking it to build buttons or to calculate the location of the galactic center. Eventually, step by step, slowly and steadily, this partnership of Matt and ChatGPT built up this application (I have it on my phone. It's great). As Matt noted, "You never want to give ChatGPT big goals where it has to figure out the way on its own. Then both of you will be confused. Intermediate stepping stones and being sure of your boots with each stride, that's the way."

To be clear, this doesn't completely eliminate your need to understand code. Software development requires effort, but prompting is also subtle and difficult. There are pros and cons to dropping into the code oneself and solving problems with a bit of doodling and intuitive understanding, as compared to iteratively conversing with generative AI to fine-tune what is being generated.

Nevertheless, this is magic. Or at least life in the world of *Star Trek*. With a series of prompts, you can eventually generate a program that points toward the supermassive black hole at the center of our galaxy. The world can dance with code, and all you need to do is ask. As AI continues to advance, such requests will become larger and easier to make, with these tools sometimes writing even entire programs at once for the user. Program synthesis allows everyday users to build programs more easily than ever before; no longer is this outsourced to an elite.

This is the goal. I don't want software creation and an understanding of computers to be "dumbed down." But I want low floors for our software as much as ceilings that are soaring and cathedralesque. Code and software are not an exclusive realm, with a clear dividing line between creator and user.

UNIVERSAL COMPUTATIONAL LITERACY

Universal literacy has been one of our modern civilizational goals, and we see its consequences everywhere, from the ubiquity of email to texting. Are these missives we write on our phones for the ages? I hope not. But even though there are different levels of writing ability, with poetry and book writing somewhere in the upper echelons, everyone can still partake of writing (and even of poetry and book writing). Literacy is egalitarian. We have to strive for the same thing with our computers.

Yes, what it means to write software will likely change. But that is both a good thing and how it has always been. I don't eke out assembly code by the sweat of my brow, and I shouldn't want us to ever return to that time.

Is the code that results from these new AI technologies going to be perfect? No. Is it going to be completely understandable? Also no. But this is the same situation with human-generated computer code. We are awash in imperfect technologies that we don't fully understand (I wrote an entire book about that). But it's incumbent on us, as humans, to at least try to grapple with the outputs of these systems and improve on them, ensuring that they operate the way the user intends them. If these tools make it a lot easier for everyone—scientists, office workers, artists, management consultants—to build software, then I will praise these systems from the rooftops.

That being said, even if code becomes generated by software, we still need to take responsibility for it. If you don't know how a piece of code works, or accept suggestions without interrogating them, something really can be lost. We pay for that loss with an abundance of glitches and failures. But as long as we are making programming the best version of itself, where it involves stripping computational thinking to its core and building powerful and useful software, then full speed ahead. In fact, working with these kinds of tools has been described as a "creative collaboration" and can help to accelerate learning. Far from stultifying us, these tools can unleash our creativity. The importance of computational thinking is less about learning the details of loops and more about the ability to break down a problem into its parts or understand the inner workings of a fundamental algorithm.

We still have a ways to go here. This kind of software creation needs to be a lot more collaborative, helping nonprogrammers figure out what they actually want in a program and turning someone's vague desires into clear ideas for how a piece of software might operate. But if technologists can bridge this gap for nonprogrammers, this realm of computational thinking will have been brought to everyone.

———

In the Middle Ages, magic was not just considered the bailiwick of sorcerers and scholars; there was everyday magic available to everyone. There was magic for finding lost objects, or helping with one's farming, or that could be used to treat illnesses or medical maladies.

Are you a farmer and you've lost your cows? Then you can perform the following: "Before anything else, call on Christ and Bethlehem. Then, look towards the east three times and say, 'The Cross of Christ is led forth from the east!' Do the same with the west, the south, and the north. To finish exclaim that 'So by this deed may nothing be hidden through the Holy Rood of Christ. Amen.'"

Whether magic or code, these are not meant to be esoteric and for the few. Code should be available to everyone. As I explored in the previous chapter, computer programs are meant to be read by humans, not just computers. And if we can do this for more people, so much the better.

In a fantastic computer advertisement from the 1980s, the programming language Logo billed itself thus: "Logo has often been described as a language for children. It is so, but in the same sense that English is a language for children, a sense that does not preclude its being ALSO a language for poets, scientists, and philosophers."

From spreadsheets and HyperCard to ChatGPT, the democratization of code and software creation is not some sideshow in the face of true programming; in many ways, it is the ultimate goal. It is a path that might even lead us to the world of *Star Trek*.

Computing as a space where you can play with software, rapidly prototype ideas, and learn about the world no matter whether you are a poet, scientist, philosopher, or child is something worth striving for.

———

We have examined the language of computational thinking— programming languages—as well as how it can be democratized. It's now time to examine how thought itself can be augmented by machines.

8

Tools for Thought

Software for Thinking

Too often, technology is at odds with humans. When I think about technology and its place in our society, I don't want it to fundamentally change us. But instead of technology helping us to become the best versions of ourselves, we often slouch toward adapting ourselves to the technologies around us. Our humanity yields to technology.

For example, I am writing this book by typing it out on a keyboard, a powerful technology with a long and winding history, with the details of the QWERTY layout lost in the mists of the late

nineteenth century. Whether the layout was designed to minimize
the bits of a typewriter snagging each other, or designed to easily
type the word "typewriter," or something else entirely, it is clear
that the specific layout that has become standard did not have to be
exactly the way it is. It was dependent on the quirks and vagaries
of individual decisions that over time have added up to what we
have now.

So we have this strange object that is the result of a specific
historical path. Everyone is now expected to communicate using
it. Becoming a proficient typist requires spending hours training to
use this machinery. In other words, we have adapted ourselves to
this piece of technology.

This is relatively minor and could just as easily be described
as learning a skill—which sounds much more positive—instead of
seeing it as bending our humanity to technology. But these user
interfaces add up. We yell at our Google Home or Amazon Alexa
in very particular ways. I use a trackpad and a mouse with my right
hand, despite being left-handed, because that is the convention
(and because when I learned how to use a mouse, I was sharing a
computer with the rest of my right-handed family). I work according to the display of a clock, rather than the rhythms of the sun.

Over and over, we have distorted ourselves to our technologies.
In many ways, it's just easier. Our relationship with work has been
altered because of the omnipresence of smartphones, with their
notifications of messages from the boss and emails from cowork-
ers. We have warped our work habits to these devices and interactions. When it comes to cognition—our ability to think deeply
and well—we have, by and large, adapted ourselves to technology
in ways that are not good for us.

Our increased drive toward the optimization of work can run
directly counter to our ability to think deeply, or to even feel that

human. Life by stopwatch or spreadsheet is not life at all. We are surrounded by the ever-present demands on our time and minds by these technologies, so that it becomes harder to concentrate, to maintain the sustained and deep attention required to generate great thoughts. Our smartphones ping us constantly, the siren calls of news and social media pull us toward the rocks of distraction, and we are always primed for the dopamine hit of a new email or like. Sitting and reading a book in this environment becomes nearly impossible.

For those of us who were using technology in the 1990s and early 2000s, pre–smartphones and social media—and the portability of our machines—distraction was less of a problem. Computers didn't feel as compelling. They were wonderful, of course—many of my digital touchstones are from this era—but they weren't addictive in quite the same way. We took walks without checking the news. We had conversations with friends without wondering about the vibrations in our pockets. We could sit still and read books.

It does not help that these new technologies, which we can take everywhere, have been engineered to attract us. Between this engineering and our own habituation, soon enough, we become used to acting like moths to these flames, whether prompted by our machines or not.

Of course, I'm not revealing anything particularly novel here. Many others have noted these trends, sounding the alarm or trying to hold back the flood.

However, I believe that when it comes to those technologies developed specifically for thinking—computational tools for thought—we can see hints of a different approach, a way of using our minds in harmony with machines, as opposed to feeling like we are their adversaries. Technology has the potential to empower us to be more deeply human and to think better. It even can involve working alongside computational thinking—that is, the world of artificial intelligence. Understanding these tools is the subject to which we now turn.

SOFTWARE FOR BETTER THINKING

We are finite creatures, frustrated by our limits, ever yearning to stretch beyond them. These bounds are confining, discomfiting our minds. We are disheartened by two limitations in particular: our reason and our time.

We cannot understand everything, read everything, remember everything. Our evolutionary trajectory has granted us fight or flight, facial recognition, and dexterity, but we are not optimized for certain kinds of abstract thought. Despite this, living in a world shaped by technologies, scientific advances, and abstract concepts, I will venture that we have done quite well. But I, at least, am painfully aware of my limitations. I cannot wade through my mind the way my fingers might dance through the contents of a filing cabinet; I must be content with helter-skelter remembrances, some reasonable, some irrelevant, and some just advertising jingles. Our powers of reason are limited. It is hard to hold complicated arguments and structures in our heads, and once we are working with large, complex systems—from biology to our own engineering works—we quickly confront our limitations. Despite the biblical tale of creation telling us that humanity has eaten of the tree of knowledge, it seems that we can do far better than this divine theft.

So, too, with time. Although Adam and Eve ate from the first Edenic tree, they were expelled before they could eat of the second tree of the garden and become even more like gods: the tree of life. Our time on this gem of a planet is limited. We are all doomed to die. We have done well with sanitation and vaccination and the sundry other gifts of modernity, but longevity and increased health do not mean living for a thousand years, or a million years, or until the stars begin to wink out.

No matter how well we think or how well we live, we cannot see or do it all. Reason and time, these twin powers of the gods, are not fully granted to us. But if we can never grasp immortality, can we make the most of the time granted to us by the slow unwinding of our mortal coils? Can we steal time back from the gods, not just for hedonistic pleasures but for improving our world and flourishing within it? In particular, can we use technologies for thinking better, making sense of the knowledge around us, grappling with the deluge of information we are cast upon?

The world of tools for thinking involves the broad class of software that tries to augment our limited reason while also grappling with our finitude, both physical and temporal. But augmenting our thought is far from a given. While computers can act as an aid to thinking, we do not necessarily view this mode as the default. We must actively strive to make these machines into tools for cognition, because so many of those in the tech world are actively bending them toward other purposes, ones diametrically opposed to deep and enriching thought.

Despite the strong current pushing us toward the cognitive shoals, we can use technology to navigate deep waters of thought. My first taste of this was with my family's first personal computer, the Commodore VIC-20. I was very young when we got it, and it was superseded by an early Macintosh several years later—one of those boxy affairs with a small oscilloscope-like black-and-white monitor—causing the VIC-20 to be exiled to basement storage. Over a decade later, when I was in high school, spelunking downstairs, I found the boxes that stored this machine, its various peripherals, cassette tapes, manuals, and other assorted detritus that any personal computer accumulates.

Digging into these boxes, I discovered that my father hadn't just been using computer programs; he had been writing them. In retrospect, this shouldn't have been particularly surprising. In order to use these early machines, you had to be at least comfortable with writing generous amounts of code. That was one major way of getting new software. The boundary between user and creator during this time was a fuzzy one. And while my father's vocation was dermatology, he had long been an early adopter of technology. He learned computer programming in high school, and briefly took a programming course in college that involved trotting around campus with a stack of punch cards and feeding them into the university's mainframe. This programming method was too onerous, though, and he quickly dropped the course.

One of the programs he had worked on for our VIC-20 was a set of educational flashcards. My father had built software specifically for me. There were texts of questions and answers and I think even rudimentary graphics that my father had been creating. While I don't remember using this program, since I had been so young, this was my first taste of a computational tool for thought, a mechanism for augmenting the memory of a child.

But there is so much more possible. The world of tools for thought is a vast one.

———

For many of us, the mind is, to borrow a phrase, a "dark illimitable ocean," unable to be fully grasped, with nothing but post hoc rationalization providing the means for understanding our thoughts and choices. But it is not actually this way. The mind is not a murky and unknowable void, with ideas and words emerging unbidden.

Scientists have slowly been shining light into this world, learning how words are stored in the mind, how memories can be processed more effectively, and even the reasons why we might feel that a word is on the tip of our tongue, just out of our mental grasp. And while the mechanism by which new ideas and concepts arise can feel mysterious, there is a method here as well. The process of generating new ideas involves recombination, inspiration, and subconscious simmering of preexisting knowledge. Cognitive scientists and psychologists study how creativity operates in the mind, and there are even university programs devoted to creative studies, articulating the determinants of creativity, the dos and don'ts of brainstorming, and the origins of new ideas.

Of course, thought is much more than these matters. It doesn't involve just ideas and words but also emotions and feelings. There are perseverations and panic, qualia and concerns, imagery and artistic sensibilities. To conjure up the image of a sunset is far more than a set of photons upon the retina, along with the metadata of where, when, and with whom you saw it. There is also the feel of the sunlight on your skin, the breeze from the ocean—for this is a sunset by the coast, after all—and the long string of emotions this evokes. The connotation of a memory is much more than specific words.

Tools of thought are not really about this, though. Perhaps to our regret, tools for thought are generally about information and making better use of it. Though this is only a small subset of what our brain does, it is an important feature, and it is becoming increasingly important as our economy becomes more infused with such work. Tools for thought don't lobotomize our thinking; they simply enhance a subset of these concerns and are able to avail themselves of the insights into how ideas are formed and how understanding can be achieved.

While I have told a tale of technology often making us less human, it turns out that some of the earliest computing pioneers saw these machines as having the potential for aiding our thinking. It is time now to examine this history.

A PREHISTORY OF TOOLS FOR THOUGHT

The modern origins of technology to improve our thinking is traditionally located in a thought experiment. In 1945, the physicist and proto–computer scientist Vannevar Bush wrote an essay for *The Atlantic* entitled "As We May Think." In it, Bush developed the idea of the memex, a vision of interconnected texts using microfilm and screens that would bring together a user's notes and ideas while capturing the winding path she might take in her research. Bush was motivated by a problem that sounds quite familiar to our modern minds: there was an increasingly huge amount of scientific knowledge to comb through, and, due to the nature of specialization, no scientist could ever hope to wade through it all. So, to address this problem of informational overwhelm, Bush basically, in a thought experiment, invented the World Wide Web, combined with one's bespoke notes hyperlinked together. This grand vision remained vaporware for several decades.

But it turns out that Bush was not the first to conceive of something along these lines. In the decades prior, Paul Otlet—a Belgian bibliographer and cataloger of books and knowledge—had worked on a series of ideas that similarly stitched together information. Otlet built a classification mechanism for books and documents, a sort of combinatorial Dewey decimal system, using numbers and symbols to show the specific nature of a book and its topics. Rather than drilling down into more and more fine-grained categories, it allowed the user to agglomerate categories. Otlet's web was called the Mundaneum—early versions were housed in the Palais Mondial—and there were even Otlet-made schematics of a sort of Belgian memex called the Mondothèque.

But one can go even further back in time. For example, we can study Emanuel Goldberg, who developed a kind of "dial-up search engine." And it was, according to Alex Wright in *Cataloging the World*, something that was actually built. The history of technology augmenting our brains is an old and long one.

This history begins—or at least let us begin it—with literacy. Without written text, information must be passed along orally or knowledge must be demonstrated physically (you can learn to make an arrowhead by watching someone do it, instead of reading a text). While an oral culture has many good features—including making the information's maintenance dependent on a thriving community, similar to what we've explored in relation to open source software—it obviously has a lot of limitations.

Without the written word, you're limited by people's memory and who is around you and what they know. But with text, you can learn from those who might be far away, or who lived before you were born, and you can now outsource your memory to writing and books.

Written text gets even fancier once methods for doing arithmetic or keeping track of data are developed. Paper—or whatever surface you're using, such as a wax tablet—is now a partner in your calculations. When your teachers admonished you to show your work, they might have been trying to understand how you arrived at your solution or searching for a way to give you partial credit, but they were also doing something else: they were ensuring that your brain's work was being augmented by a tool.

Related to this, there were advances in notation, from the use of the number zero to the Arabic numeral system, which is the decimal system that we use today. (I, for one, am glad that we discarded the Roman numeral system, which—and maybe I'm being unfair here, as I am not a native user of it—is basically insane.)

As history proceeded, we experienced a steady march of so many more tools for thought. There was the advent of the codex—a book with pages, instead of a scroll—which allowed for random access to information contained in a text. There was the combined compass and ruler, which allowed geometers to do their thing and generate shapes and constructions. There was the printing press, allowing knowledge to be easily spread and copied with high fidelity; we no longer had to rely on monks whose attention might wander while they transcribed manuscripts. There were different writing tools and surfaces: better and cheaper paper, the pencil, the eraser, the eraser on the end of the pencil, and the ballpoint pen to avoid smudging and messiness.

There was the invention of the index at the end of books, allowing texts to be accessed nonlinearly, skipping from page to page. (A well-crafted index is a joy to behold: it contains the essence of the book, addressing the idiosyncratic ways that we each organize information in our own minds and think about topics and subjects.) There were the dueling calculus notations of Isaac Newton and Gottfried Wilhelm Leibniz, both of which are still in use in different situations. There is the concept of alphabetization itself—organizing information in alphabetical order was far from a given even as late as several centuries ago—and the commonplace book, which collected the various quotations someone had come across.

And then, we finally had the note card.

FROM THE INDEX CARD TO
HUMAN-MACHINE SYMBIOSIS

The shortest-lived NFL team in the league's history played one game and lost badly, forty-five to zero. Known as the Tonawanda Lumbermen or Tonawanda Kardex, the team was based in Tonawanda, a town in the suburbs of Buffalo, not that far from where I grew up. I suppose this might be a point of pride for the city, though I imagine many will disagree. While the team existed prior to its inauspicious 1921 NFL performance, it was only in the league for one game. But buried deep in this bit of history and lore is the story behind this football team's weird name: Kardex.

American Kardex was a company founded in 1915—and still around, through its multiple acquisitions and mergers—devoted to information management and office supplies. It was based in Tonawanda, and its flagship product was a kind of index card and associated filing cabinet known as a Kardex. For a time, it was so popular that it appears to have been used in some places as a generic term for such cards in the medical profession, helpful for note keeping and storing health records.

Index cards were the technology of the future. While cards for storing and managing information had been used for centuries, only in the nineteenth century did this technology begin to be standardized and organized. For example, Paul Otlet of the Mundaneum and the Mondothèque—whom we've already met—was a connoisseur, with his Mundaneum employing millions of index cards according to his particular system of notation known as the Universal Decimal Classification.

Index cards, the codex, pencils, and erasers are simple gadgets. But soon enough, machinery began to enter the realm of tools for thought. The typewriter was an important step here, allowing a faster mechanism for conveying the thoughts in your head onto a sheet of paper. This soon gave way to word processors, special-purpose electronic machines for doing nothing more than writing text (there is no doubt a certain amount of envy here in the modern writer: a machine absent any distractions from the Internet, doing nothing but letting you write). These word processors, which allowed for text to be stored and had the potential for editing, were built by companies with names that did not exactly roll off the tongue: Linolex, Lexitron, Wang, Redactron, Vydec.

Finally, finally, the properties of the word processor and the storage medium of the index card came together in the form of the digital personal computer. But even when the digital computer was a room-scale behemoth, requiring sprawling cooling systems and constant vigilance to maintain its operation, it was a tool for thought. It allowed the calculation of things that could be described but not achieved by humans. I can add any two numbers together or multiply any two numbers. Can I do this well, or fast, or accurately? Not always. For big numbers, I am almost guaranteed to make a mistake. Can I do this one hundred thousand times in a row? Theoretically yes, but practically? No.

In other words, I can describe a task of the mind and then set a computer to do it, even though I cannot accomplish it myself. And that's a tool for thought. Whether it's calculating ballistic trajectories for the military, or making weather predictions for a whole country, or creating mathematical worlds, these computers allowed us to take the shape of calculations and bring them to a result. But even more than that, computers provided textual and graphical ways of manipulating information, from navigating entire corpora of documents to streams of numbers. A core use of the computer is as a tool for thinking.

The computing era allowed for a profusion of such tools and approaches. There is a kind of pantheon of individuals involved in this space, a specific procession of names: J. C. R. Licklider, Ivan Sutherland, Douglas Engelbart, Alan Kay, Adele Goldberg, Ted Nelson. At the risk of avoiding the weeds and skipping over something essential, these are the parents of intelligence augmentation through the use of computers.

J. C. R. Licklider was originally trained as a psychologist until he became entranced by the possibilities of computers, and he was responsible for funding much of the research in this area through his work at the Department of Defense Advanced Research Projects Agency. Licklider articulated his vision for what was possible with these machines in a 1960 paper entitled "Man-Computer Symbiosis." In this paper, Lick—as he was known—described his idea of symbiosis as something distinct from machines entirely replacing human thought. Licklider recognized that eventually artificial intelligence might very well have its day, but in the meantime, for however long that is, we must work to construct a sort of symbiotic relationship between humans and machines, where machines make humans better, and humans improve the information-processing of computers. So, Lick studied his research process and began to examine which of his own activities might be capable of being done by machines. He found that computers could help with the tasks of searching, transforming and visualizing information, and much more. Lick anticipated the screens and interfaces that would be needed, though he missed the mark when it came to the amount of information that could be stored in our computers. After estimating that the technical literature would require billions of dollars in funding to be stored, he noted that "the first thing to face is that we shall not store all the technical and scientific papers in computer memory." Nevertheless, Licklider recognized the importance of allowing humans to do what we do best and machines to do what they do best.

Only a couple years later, Ivan Sutherland, knowing of both Licklider and Vannevar Bush, set out to build something that embodied these ideas in software, which he called Sketchpad. Described in a report in 1963, Sketchpad used a kind of stylus to manipulate information and shapes on a screen, intuitively and easily, particularly for technical drawings. Sutherland recognized that text, while a potent communication medium, is far from universal. Being able to manipulate shapes directly can be incredibly powerful. For example, Sketchpad could be used to find the forces involved in the design of a bridge, entirely on the machine. This feedback between design and information is a powerful enhancer.

Douglas Engelbart is primarily known for "the Mother of All Demos," an actual demonstration and talk, where over the course of ninety minutes or so, he showed off a word processor, a computer mouse, hyperlinks, and even video conferencing. This was all in 1968, and it has boggled minds ever since. Several years earlier, Engelbart also wrote a report on "Augmenting Human Intellect," exploring the ways our tools affect our thinking and how computers can augment our abilities.

Alan Kay and Adele Goldberg were involved with work at Xerox PARC in the 1970s that led to the personal computer and modern graphical interface (the desktop metaphor, windows, and all that). They also developed the idea of the Dynabook, a sort of iPad-laptop hybrid geared toward children first thought of by Kay. Ted Nelson coined the terms "hyperlink" and "hypertext" and pushed for the idea of interconnected information.

From these people and many more like them came the ideas of the modern personal computing revolution, and in particular how we think about how computers augment our ability to think and learn.

Sadly, the work of these people is also one of various kinds of disappointment: disappointment in the disjunction between their vision and reality, or between creation and lack of adoption, or simply invention alongside a kind of societal amnesia. Unless you are steeped in this world, these names are not ones that linger on the lips; these people have been forgotten or were never even learned about by nearly everyone who touches a computer. But ultimately, much of the disappointment here might simply be the disappointment of a prophet in his people: I see so clearly the future and the promised land, and yet, over and over, you have disappointed me.

The Dynabook wasn't created. Ted Nelson, in addition to being known for hypertext, is also associated with a system known as Xanadu, which is notorious for having been under development for decades. And yet, while the Xerox Alto was never really a commercial device—this was a computer that incorporated many of the foundational ideas of these people—it did influence basically the entire modern computing world, from the Apple Macintosh to the Windows operating system, to the development of Ethernet, HyperCard, Microsoft Word, the PDF format, and modern email clients. All of these ideas have permeated the world of computation.

We now, at last, come to the combinatorial digital tool to rule them all, the one designed to stitch together all knowledge, growing haphazardly and piecemeal, full of wonders and nonsense: the World Wide Web. Combining the ideas of hypertext with a distributed structure, the Web finally cashed the promissory notes of Bush and Otlet, Licklider and Nelson. We now live in their future, finally able to lay down Bush's "associative trails," mimicking the way that our minds process information and think about concepts.

With the founding era of interfaces, frameworks, and technologies all layered on top of this information crisscrossing the globe, these ideas have cross-reacted, recombined, and resulted in a vast burst of dandelion pollen. Today we are awash in tools for thinking of every kind.

Tools for arranging notes and stitching together one's half-baked ideas abound. Software that allows for greater intuition around numbers and quantities has blossomed. And there are tools for visualizing information. Many of these applications aim to provide a playfulness and a way of manipulating information and ideas, sometimes directly and sometimes indirectly. There are tools for remembering better. Some of these take advantage of technology far beyond the flashcard, making use of spaced repetition, which shows you something right when you are most likely to forget it in order to make certain it is always remembered. There is software that teaches in a dynamic way, through learning by doing. Spreadsheets and other simulation tools allow models of our world to be constructed, which can help us think better about these complex systems, something we will see more in the next chapter. There are even computer languages designed for better thinking, such as Logo. This is a world steeped in education and how children learn, or the study of how the mind makes connections, memorizes, and retains information. It embraces the idea that thinking is something that humans are good at but can do better.

We even see hints of fantasy made real. In the *Harry Potter* series, there is a device known as the Pensieve. A large stone basin, it is described as a place for storing thoughts and memories when you simply have too much to think about. This is the impetus for many tools of thought, from index cards, to software that collects and connects text notes, to the very act of archiving your email. Magic and wisps of silvery thought swirled into a stone bowl are made real by computation.

Intelligence itself—also long the province of our stories—is now becoming the realm of machines.

ARTIFICIAL THINKING

What is artificial intelligence? AI has already lurked at the edges of many of the topics I've explored so far, but it is time to bring it out of the shadows, from a tool or add-on to a force in its own right. Simply put, artificial intelligence is a loose bundle of techniques and approaches that enable computers to do things that might be considered to involve thinking or resemble human intelligence in a range of areas. Or, more cynically, it's anything that computers can't do yet. For AI has long seemed to be one of those moving targets; if a machine can do it, it can't possibly require intelligence. Can a computer play tic-tac-toe or chess? Well, maybe true intelligence wasn't involved. Can it compose high school essays? Again, maybe that doesn't require true intelligence. One of the founders of artificial intelligence, John McCarthy, bemoaned the fact that "as soon as it works, no one calls it AI anymore."

Regardless, the field of AI encompasses robotics, generating text and images, recommendation systems found in online retailers, transcribing speech, game playing, and all manner of other activities. It first emerged as a domain of study in 1956, when a group of luminaries got together at a conference at Dartmouth College, including McCarthy; Marvin Minsky, another eventual giant of AI; and Claude Shannon, who invented information theory. Most of the tasks of AI—using language, problem-solving, and the like—were initially thought to be so tractable that they would take a summer or so to make headway on.

This was not correct.

In the ensuing decades, through a combination of more sophisticated algorithms, overflowing databases, and enough computing power to make a reader of *Creative Computing* magazine weep, AI became a reality. Between then and now there were hype cycles, booms and busts, promises broken. But in the Twenty-First Century—capitalized, like it's been pulled from a golden age science fiction paperback—AI went from brittle and trivial to robust and general-purpose. Image recognition that challenged human abilities was achieved in 2015, and, as mentioned in Chapter 1, the Turing Test—the ability to have a chat conversation and not know if it's a person or a machine—seems to have been blown past sometime in late 2022. (Where were the parades? The balloon drops? I was denied this milestone, but that is a separate matter.)

How were most of these recent advances in AI made? Through the use of neural networks. Neural networks are what they sound like: a network of digital neurons all wired up together into a big old web. But unlike the neurons in our brains, these are described in software and are just straightforward mathematical operations: take input values, apply some arithmetic, and then output the result (such as a number between 0 and 1). When you connect enough of these rudimentary artificial neurons and allow the results of the output to be grist for tuning these arithmetic functions—this is the so-called training of the neural network—then eventually these networks begin to do some amazing things. This is the magic of algorithm, data, and computing power: with the right algorithm and enough data and compute—for the training is a computationally intensive optimization process—then almost anything is possible. For example, the most recent AI systems in wide use, such as ChatGPT, have parameters in the many billions within their neural networks, these parameters being the means of tuning the specifics of the arithmetic functions in their neurons that result in their "intelligent" output.

There are many different architectures of these neural networks, with details involving everything from how they receive their input to how many layers of neurons a network might have. But when it comes to grappling with vast quantities of knowledge—specifically textual knowledge—a particular network architecture known as the transformer was devised to manage it all. This is the T in the GPT models of OpenAI, and it works well because, in part, it can be scaled up and process everything in parallel. Transformers are ultimately involved in prediction: trying to determine the next word (or really a "token," which is a bit different from a single word, but thinking about words is good enough) for a certain text—essentially continuing it—based on the statistical properties of language. This is a handy trick if you want to generate new text.

An entirely rudimentary and naive model of language prediction would involve figuring out the likelihoods of all the words in English and rolling a virtual many-sided die, weighted according to the probability for each new word. "The" would come up a lot, "syzygy" not so much. A string of die rolls would give you a sentence based on the frequency of words. But, of course, we don't just utter words based on how common they are. That's not how language works. If I began a sentence with "Once upon a," it's not very likely to be "Once upon a nightstand," no matter how much we like to talk about furniture.

This means that we need to incorporate the likelihood of a word based on the words that have come before it. But aside from common phrases, most possible strings of a few words or more won't exist in written language we find in the wild. So what to do? Neural networks can be used to create an internal representation of the patterns and foibles of English—or any other language—in order to predict what word will come next.

To do this, transformers contain within them something known as an embedding space, essentially a mathematical space of meanings (though these are not unique to transformers). Transformers place text into this space and then turn it back into generated text.

An embedding is essentially a way to mathematically represent a word or chunk of text. But instead of storing it as text, or even as some string in binary, the program stores it as a location—a series of coordinates—in a compressed, many-dimensional space. Or, more properly, text is stored as a vector: an arrow pointing toward a specific point in this space of many dimensions. Instead of each coordinate for this vector denoting height, width, depth, or some other higher spatial dimension to which our minds have no access, each dimension corresponds to a vague semantic property. This is because the locations of words in this space—where they are "embedded"—is based on the words that are often found nearby in texts. While this sounds either trivial or uninformative, as long as you pour in a large enough corpus of text, you end up getting out a certain amount of meaning. For example, there is probably something like tens of trillions of words available on the public Web.

The clearest way to see the power of these embeddings is through some word math. What does *king* – *man* + *woman* equal? In the embedding space, this is math you can actually do. As per this well-known example, *king* – *man* + *woman* = *queen* (at least approximately). Once you subtract the idea of masculinity from the concept of royalty but then add back in a lady, you get a queen. And this is what these kinds of embeddings give us. *Celebration* – *fun* = *commemoration*. Or even *Hitler* – *Germany* + *Italy* = *Mussolini*. In the large space of words and phrases, their locations and relative positions correspond to intuitions we have about the meanings of these words. Older AI systems might have hardwired the logic, syntax, and meaning of words and how language works, but in this case, neural networks—when trained on enough information—allow these meanings and features to simply percolate to the surface from this combinatorial and indecipherable process. Meaning emerges.

The fact that meaning emerges from this mess of text—which can be used to better think about the messiness of language itself—is something that might be intuitive after the fact, but, to me at least, it is more than a bit surprising. What this means is that the fuzzy nature of meaning needn't be derived from a dictionary; it's simply inherent in our use of language. And this meaning can be employed to better understand how we use language and as a foundation for generating new text that is both logical and comprehensible.

Transformers operate by looking at large chunks of text at a time, using the same ideas as embeddings but also incorporating the context in which a word appears. Instead of just having a single embedding for any particular word, each word's placement in this coordinate space can be refined by the context of all the words that came before it in the inputted text, which provides a better sense of that word's meaning. This, then, generates a predicted next word. And thus, ChatGPT!

But not really. There are many specific details and features of these systems I have not discussed. For example, there is a setting known as "temperature," which determines how often the system chooses a less likely word: the higher the temperature, the weirder or more creative the output. There are also many details of how it is designed to operate like a chatbot that I skipped over. But in the end, these large models, through their embeddings both simple and complex, create a multidimensional "latent space"—a space of meanings, some clear and some not, that dictate how these models crunch text.

This space of meaning, though, must be handled with caution. The world of large, generative neural network models is one based on stories and text, not on physical models. Sometimes they overlap, but they needn't always do so. So we get hallucinations or strangeness of all kinds out of these GPT-style models, from recommendations to use glue when making pizza to prevent the cheese from slipping off to references to imaginary scientific papers that don't exist.

Nevertheless, from these crooked timbers we can build a great deal. AI can assist us in grappling with vast amounts of information. When there is too much for any single person to read and understand, we need computers to help us. Recall that Vannevar Bush's memex was motivated by the fact that there was an increasingly huge amount of scientific knowledge to comb through and that, due to specialization, no scientist could ever hope to wade through it all.

These massive networks have encoded within them the way that language is used, at least based on the body of text they were trained on (so there are potential biases here, of course). Therefore, they can be used—even in conjunction with other techniques—to navigate large tracts of information. Because location and distance are based on meaning, scientific papers, for example, can be placed into this space and then navigated more easily. Researchers have even claimed that by using such techniques, certain discoveries might have been made years earlier. For example, by using data from materials science only up to 2009, a certain chemical compound known as $CuGaTe_2$ could have been hypothesized to have thermoelectric properties—converting between heat and electricity—a few years before a paper with evidence of this was published. When the researchers placed this chemical's term into the latent space of the literature, they found that it was very close to the word "thermoelectric," even though it had not been known to have this property.

From AI leaps so much. Summarization tools: rapidly learn an entire emerging field! Recommendation tools: if you liked this paper or nugget of information, you might like this! Or even tools for generating new ideas, from hypotheses to product names. AI can provide a kind of engineered serendipity or even "augmented imagination."

Generative AI's reliance on constructing a complex and navigable space between words and meanings has great potential in helping us to think better. As well it should, because AI tools are essentially based on landscapes of thought and information: the latent spaces that embed text are maps for this terrain. AI is therefore particularly well positioned to, for example, find information that could be relevant to whatever you might be working on or studying at the moment. This kind of surfacing is a precondition for better understanding a topic and for synthesizing separate concepts to develop something new. What you might do now in an imperfect and meandering way—reading one article and being led to another, or doing some keyword searches online—has the potential for being done more systematically. Even education itself—a task bursting with thought—is an activity that is amenable to the use of AI, from helping find new sources for study to even just acting as a powerful teacher you can ask question after question of as you learn something new, never feeling judged for a silly query.

The latent spaces within AI models offer maps to the high-dimensional topography of information, and we can use them to augment our thinking.

TO BE HUMAN

For some, life is absurd. Life, as per Macbeth, is a "tale told by an idiot, full of sound and fury, signifying nothing." Or if you prefer *Seinfeld*, when discussing why he doesn't like cliff-hanger television episodes, Jerry Seinfeld noted the following: "If I wanted a long, boring story with no point to it, I have my life." And yet, meaning still emerges out of the nihilistic raw material of the facts of our lives. It is something that we have to search for, whether yielding a limited and humble sort of meaning or something that feels like a dent in the cosmos.

But when we think of computing alongside this search for human flourishing, we get uneasy. And rightly so. As I noted at the outset of this chapter, slouching toward mechanization is the path of least resistance. But these tools for thought provide a hint of other possible ends. For example, the poet and novelist Richard Brautigan, an icon of the 1960s counterculture and proponent of connecting with nature, provided a different—and very sixties—vision of the future. In his 1967 poem, "All Watched Over by Machines of Loving Grace," Brautigan depicted an idyllic future of "a cybernetic meadow / where mammals and computers / live together in mutually / programming harmony." While naive to our modern ears, it certainly seems a better outcome than other, more dehumanized, possibilities.

As artificial intelligence continues its march forward, there is a clear path toward an outsourcing of our thinking, our reason. We might have stolen the fruit of the tree of knowledge, but for what? To give it up to machines? Generative AI, like ChatGPT, can now do things that were long thought to be the province of humanity. If these abilities are no longer unique to human beings, where does that leave us?

This sort of worry is a constantly shifting baseline. Some might have once thought tool use was the sole province of humans. But we now know that many other members of the animal kingdom use tools. They also use language of sorts (apes and dolphins are proficient) and have the concept of transmissible culture (birds have this as well).

So, too, is it with every advance in AI: checkers, chess, art, mediocre poetry—all are falling to machines. These are not the exclusive domain of humanity.

These changes over time can cause a certain amount of confusion and cognitive dissonance. This happened to the cognitive scientist Douglas Hofstadter. Hofstadter is the author of *Gödel, Escher, Bach: An Eternal Golden Braid* and *I Am a Strange Loop*, seminal books that sought to explain the origins of thought and mind and their connections to computation. As the author Brian Christian noted in *The Most Human Human*—an exploration of artificial intelligence and the story of Christian's own participation in a kind of Turing Test—Hofstadter had argued that chess required sophisticated thought and subtle intelligence. Until, that is, Deep Blue and Garry Kasparov. After their first chess match in 1996, Hofstadter moved the goalposts: "My God, I used to think chess required thought. Now, I realize it doesn't." While humans might need to think for chess, computers can thump humans at the game without any true thought; they can just calculate.

This is reminiscent of "God of the gaps" arguments in theology: What is God? God is everything that is not explained by science. But then we have the advent of evolution, astrophysics, cosmology, and neuroscience. Suddenly God doesn't seem so big anymore, defined into nonexistence by the march of progress. Whether or not a specific task requires thinking to be accomplished by machines, AI is playing the same role of science here, sweeping away vast swaths of what we had believed were unique domains of human thought.

Don't make the same mistake as these theologians when it comes to AI and our humanity, defining uniqueness—whether of God or human thought—as something smaller and smaller. This uniqueness might all go away. Focus instead on our quintessential humanity, what we care most about actually doing, whether or not AI can do it as well. There are many answers to what that might be, and it is for each of us to figure out. But at least one aspect of this quintessential humanity that I'd like to proffer is our ability to think and reason. In the words of the prolific online essayist Venkatesh Rao, thought is—or at least should be—autotelic: thinking is good as its own end. It's fun to think, to roll ideas around in our minds and develop them. There is a pleasure to thought. So instead of trying to outsource it to machines—even if they can do it well—we must strive to make it easier and more enjoyable to do. We can eliminate some of the friction in coming up with novel concepts or remove some of the frustration in connecting ideas together. We now have the ability to use what are referred to as "epistemic technologies" to assist in knowledge work, distinct from so-called production technologies like automated manufacturing. Sometimes these epistemic technologies will involve AI—suggesting ideas to recombine or working through concepts with you—but sometimes they might involve using far older technologies in conjunction with computing.

For example, Andrew Sutherland, the founder of Quizlet—a popular tool for making digital flashcards that he founded when he was fifteen years old—has actually been thinking about this deeply for some time. He has been trying to blend our computers with a technology that promotes a slower and more deliberate pace of thinking: paper. Sutherland prints his emails out, handwrites responses, and then scans them back in. He even prints articles and leaves them strewn about his workspace for when they might catch his attention. Tools for thought are not all AI and outsourcing thinking; they come in all shapes and modes.

Steve Jobs rhapsodized about the transformative power of personal computers, as noted earlier, with his evocative phrase "bicycles for the mind." While humans have disadvantages—we don't do well in extremely cold and hot environments, and we lack sharp teeth and nasty claws—things change for our mobility when we hop onto a bicycle. Bicycles allow us to use energy efficiently to travel quickly where we want to go. This is what Jobs meant by computers being bicycles for the mind. They are machines that allow our innate powers to be supercharged and overclocked. This must be the goal of computing, particularly when it comes to thought: not to replace us or to narrow our humanity but to make us better versions of ourselves.

Ivan Illich was a Catholic priest turned philosopher active primarily during the 1970s and 1980s. He described tools that are aligned with our humanity as "convivial" tools, and he advocated for them. Conversely, the novelist, poet, and environmentalist Wendell Berry—who still maintains a farm in Kentucky—warned, "It is easy for me to imagine that the next great division of the world will be between people who wish to live as creatures and people who wish to live as machines." We must choose to live as creatures, and decide what we wish to be as humans.

Computers are indeed bicycles for the mind. And they have become unimaginably powerful compared to where they started decades ago. But we must remember that they are not the only way to get around.

PART III

Reality

9

Let There Be Numerical Modeling

The Worlds of Simulation

Human history and mythology are awash in creation stories. A slain frost giant provides the raw materials for existence; the world's creatures come from a single egg; a god creates the world using a pigeon, a chicken, and a snail shell full of magical earth. But probably the best-known creation story comes from the Bible. And it uses words. It is speech that created the universe: Let there be everything. And it was good.

Building worlds with text is something that we are now able to do. We can encapsulate the world around us—reality itself—within computer models, setting this description into motion within our machines. Words in myth that create worlds and the process of computational simulation begin to seem not so different.

As impressive as this advance is, there is peril in equating a creation tale with computational simulation. These models within our computers are powerful indeed, and they allow us to better grasp the world, but they are not the real thing. There is a gap between reality and computational creation. We partly forget the difference between the world and its digital version because we view many such models as nothing more than tools for prediction. They can often be this, but models are also for explanation and understanding.

Only by recognizing the nature of simulation itself—what it's good for, and what it's not—can we proceed with humility, or at least caution, rather than hubris. If we honor models' limits, balancing complexity with accuracy, they can teach us something about our surroundings. For despite harnessing massive computational resources, they are not the real world.

We can begin to understand this by examining the history of computational simulation, how we became able to build these digital worlds, as well as the power and pitfalls in their creation.

THE ORIGINS OF DIGITAL WORLDS

Creation stories have been ingrained in our collective psyches for millennia. While our models of computational simulation are of more recent vintage, we have been building models—partial instantiations of these creation myths—for thousands of years.

Prior to digital computers, models might be divided into two types: inert descriptions and physical models. For example, mathematical equations are powerful but inert, lacking life or dynamics. However insightful an equation might be, it just sits on the page, staring at you, daring you to extract its insights.

Physical models, though, can be played with. The most rudimentary do not move of their own accord. Think, for example, of a dollhouse, where the only dynamics are those made by the person adjusting the location of the furniture or moving the dolls throughout its rooms. Or a globe, allowing you to hold the earth in your hands and see continents, cities, oceans, and their relative positions. But there are physical models that also incorporate more complicated relationships and movement between the components. One well-known type is an orrery, a mechanical model of the solar system, or at least a subset of it, which allows the user to crank away, literally, and see how the moon and earth rotate and revolve, leading to the months and seasons and years. The gear ratios of the orrery are precisely tuned to show the movements of the celestial bodies around the sun, and as you crank, you can experience everything from the phases of the moon to solar eclipses.

As orreries and astrolabes—devices for calculating the positions of celestial objects and determining the time—became more sophisticated, they shaded into the world of analog computing. Analog computing can be viewed as a kind of precursor to modern digital computers—though analog computers were still used for decades after the advent of digital ones—and is vital for thinking about models and simulations of reality.

What is an analog computer? When people think of something as "analog," it is often viewed as having smooth, continuous change, as opposed to digital, where there are stepwise changes. (Some might even use analog as a kind of synonym for "retro.") For example, an analog vinyl music record encodes sound as wobbling grooves in a disc, versus a digital MP3 file, where it's all ones and zeros. That is part of it, at least generally (there are exceptions). But analog computers are not just about continuous change. They are, in fact, what it says on the tin: analogies.

An analog computer is an analogy to some physical process. A gear's position is analogous to the rotation of the earth, as in an orrery, or the height of water in a tube is analogous to the amount of money in the economy. It is computing by analogy, something incredibly useful when thinking about models.

There actually was an economic water machine, the MONIAC, and it is an example of an analog computer. It used pipes and tanks to mimic the way that money flows through a country. In the process, it calculated various values that an aquatic economist might wish to understand. An impressive feature of this machine, developed in 1949, was that the values it calculated could be easily seen. Look at the height of the water in a tube or a tank and you could immediately understand the effect of a change in tax policy.

One of the most sophisticated analog computers was made by Vannevar Bush, whose memex was mentioned earlier. Bush's analog computer, the differential analyzer, was a series of wheels and discs combined with electrical circuits that interacted to allow for the solution of differential equations. These differential equations were essentially rates of change for different variables, the properties of the system being studied. You took whatever system you were examining, determined the relevant equations that described this system, and then you could program this machine with these equations, building your particular system of mathematics from the mechanical and electrical interactions in the differential analyzer. These machines were powerful examples of analog computing, until digital computing eventually swept most of this away.

Digital computers proved so powerful because they were easily programmable and open-ended in how they could be used. While often very fast, analog computers required constructing something out of physics—a specific type of gear or the resistance in a wire—that allowed a relevant analogy to be demonstrated. You had to find a specific physical property of the world that corresponded to whatever you were trying to study; you had to search for the right physical analogy. But then digital computers arrived, able to be programmed with text instead of analogy. You could now describe a model through a set of equations or rules, embody them in code, and then turn it into a simulation.

The two modes of modeling had at last come together in these open-ended digital computers: we now were able to have inert descriptions that could come alive when run in machines.

Reality was encapsulated in code almost immediately from the moment this became possible. Beginning around the end of World War II, large digital computers—programmed with the rules of a virtual world and the necessary information collected from reality—were used to model how weather operated and how ballistic missiles found their targets. One of the first uses of the earliest digital computers was to model nuclear bombs. Computers were machines designed to slurp up and digest data, and a lot of that numerical mastication was done for prediction purposes, from weapons to weather.

Several decades after the advent of digital computers and large-scale modeling, a computer simulation called World3 captured the public consciousness. Outlined in the immensely popular book *The Limits to Growth* and published in conjunction with the Club of Rome—a group devoted to exploring how to solve global systemic issues—World3 attempted to make a computational model of the entire globe and understand how our policies and collective decisions would affect the trajectory of human civilization. Based on the field of system dynamics and published in 1972, *The Limits to Growth* was a tour de force of modeling, incorporating demographics, agriculture, land usage, industrial output, material resources of the planet, and more. World3 simulated the coming decades based on a few different global scenarios, from maintaining current trends as they were, to changing how we manage pollution, to even altering the amount of resources that would be available to humanity.

The model operated by connecting different quantities, with the result looking like a massive series of looping interconnections, a tangled web of dependencies. For example, population growth was understood as the combined total of births and deaths—something not particularly controversial—but the number of births and the number of deaths were in turn dependent on access to birth control, food production, and the like. Each of these values could be dependent on yet more quantities, with feedback—sometimes delayed—tying everything together. A graphical version of the model is splayed across two pages of *The Limits to Growth*, with the features and text so small that I struggle to read its details. Though they had built a system of great complexity, the authors take pains to mention the simplifications that they had to make in order to develop this model.

And it was a massive phenomenon. The book sold millions of copies and was translated into at least thirty languages.

But computational power—along with modeling abilities—has come far since World3 half a century ago. I can now easily download a version of this model as a Python package, and it's available as a Mathematica function. I can even just run it in my web browser and unfurl decades of the planet's future in seconds.

We now have the ability to simulate worlds inside our machines. Our computers have become ever faster and more powerful. But are there limits to what we can know?

LAPLACE AND LOVECRAFT

Prior to our era of computational simulation, the many different ways that people thought about the complexity of the world might be reducible down to two viewpoints, embodied within two individuals: Laplace and Lovecraft. These two thinkers—the former a physicist from the eighteenth and nineteenth centuries who worked to apply the laws of Newtonian mechanics to understanding our solar system, and the latter a writer from the nineteenth and early twentieth centuries who composed existentially frightening eldritch tales—encompass two opposite views of the comprehensibility of our world.

The scientist Pierre-Simon Laplace is well-known for embracing a clockwork universe subject to clear and understandable laws, rules of physics that could be entirely mapped and modeled. Laplace contended that if we could know the position and velocity of all particles, along with the rules for how they interact, we would be done. All we would need to do is simply set these equations running, and the ticking of the cosmos over time—from the motion of the planets to the details within our cells—would be predicted. It was simply a matter of details, data, and proper formulas.

On the other hand, the horror writer H. P. Lovecraft held that we barely understand the thinnest veneer of an unfathomable universe. His surname has even become an adjective, Lovecraftian, to evoke this mysterious and frightening view of reality. Lovecraft was certain that to understand all of the world was folly. Humans are unable to fully do so, through and through. In the opening lines of his short story "The Call of Cthulhu," he makes this clear:

> The most merciful thing in the world, I think, is the inability of the human mind to correlate all its contents. We live on a placid island of ignorance in the midst of black seas of infinity, and it was not meant that we should voyage far. The sciences, each straining in its own direction, have hitherto harmed us little; but some day the piecing together of dissociated knowledge will open up such terrifying vistas of reality, and of our frightful position therein, that we shall either go mad from the revelation or flee from the deadly light into the peace and safety of a new dark age.

Lovecraft implies that to truly predict the world is far from our grasp. And the day we do so will only show us the path toward madness.

Setting Lovecraft's preoccupation with madness aside, this idea that we are far from understanding the laws of our universe is a compelling one. Laplace might have thought that we knew all the rules of our clockwork cosmos, but he lived before quantum mechanics and general relativity. To even attempt an understanding of our universe without these theories would have led us completely astray. Therefore, to recognize that we might not be at complete understanding is a perspective born of humility and wisdom. Yet, the sciences have made great strides. Can we simply "shut up and calculate" in a reductionist paradise? Or is it all stabs in the dark, simplifying assumptions all the way down?

It should surprise no one who has contended with dogma and reasonableness to find that the truth is in fact somewhere between these two viewpoints: obviously science has made great strides, but we are often still limited in what we can know about complex systems, even if we know their rules entirely.

———

As science has blossomed in its attempts to understand the complex systems around us, we have discovered limits. The study of fractals, chaos, and nonlinear dynamics threw a wrench into the whole "collect data and understand the entire world" favored by Laplace. Specifically, chaos theory has demonstrated that systems more complicated than a swinging pendulum can make any computational simulation rapidly diverge over time, despite minute differences in the initial conditions. Start with a tiny rounding error, or a slight difference in the original state of the system, and there's no guarantee that a prediction will be anywhere close to where reality actually finds itself.

The nature of chaos—mathematically precise chaos, not the inchoate turmoil of the universe—was first elucidated in the 1960s by Edward Lorenz in the context of weather models. Lorenz, a meteorologist, was working with a relatively simple set of equations that described how air moves through the atmosphere. After one computer run of these equations, Lorenz decided to run the model again, but starting from values that described the simulation after it had already been running—so he wouldn't have to run the entire thing from scratch—only to discover that the final results were entirely different. This led down a brief debugging rabbit hole, wherein he determined that the numbers on his printouts had been truncated slightly: the printout only showed the numbers to three places past the decimal point, rather than all six digits past the decimal, which was how the computer was actually storing its own values. Surprisingly, this tiny difference in initial conditions ended up yielding entirely different outputs after the model had been run for a certain amount of time.

This is the essence of chaos. Small differences can matter far more than we might expect, and in a way that is hard to predict. This is the butterfly effect, where the flap of a butterfly's wing somewhere on the planet might alter the weather halfway across the globe. To be clear, the values spit out by chaotic models do end up residing on a complex shape in the numerical description of this model known as a strange attractor. The original one described by Lorenz fittingly looks like a butterfly, though others can be boomerang-like or entirely different shapes. So, there is a limit to what will be unknown and there is a certain amount of order here, or at least bounds on our uncertainty. Nevertheless, where on this strange attractor shape the values will be is uncertain.

We see this phenomenon in our own modern computational weather models: beyond a certain number of days, these models are swamped by measurement errors and butterfly effects. Yet our weather models have improved. Although some people still besmirch weather prediction, slowly and steadily, through the confluence of better algorithms, better data to be poured into these programs, and more powerful computers, our weather predictions—which often combine multiple different models—have beaten back the tide of chaos. For example, in 2023, the accuracy of our models five days out was where we were for two days out twenty-five years ago.

But these weather models are very complex. The main way that they operate is that you divide the region under question—which can even be the entire globe, because everything is connected—into as fine a grid as is feasible, given the computational power at your disposal. Then you run equations for how the air in each one changes and interacts with its neighboring members of the grid. You do this over and over, faster and faster, until you get a weather prediction. If you ever were concerned that a mathematical model was too simple, study weather models and you will never be concerned again.

These models chop up the planet into tiny squares on the order of a few kilometers, can have over one hundred layers of atmosphere, and incorporate tons of data from weather measurement.

Surprisingly, one of the earliest versions of this was done before digital computers even existed. In the first quarter of the twentieth century, the mathematician and physicist Lewis Fry Richardson manually, as a sort of proof of concept, spent many weeks over the course of multiple years trying to predict a few-hours-long change in weather. In other words, the prediction took far longer to make than the weather itself took to form. Imagine computing tomorrow's weather forecast, but only learning about it a week from today, or next year. You could figure out whether the model is accurate—sadly, Richardson's prediction was way off—but for actual prediction it would be entirely useless. Even an early digital computer weather forecast wouldn't have been timely: a weather prediction from ENIAC, one of the early digital computers, took a little more than twenty-four hours to calculate the weather twenty-four hours in advance. But all of this was solved by advances in computing, including Moore's law, which elaborates the steady doubling of processing power every couple of years. And indeed, we have come far from Richardson's days, which was of course his dream: "Perhaps some day in the dim future it will be possible to advance the computations faster than the weather advances and at a cost less than the saving to mankind due to the information gained. But that is a dream."

Weather prediction has used more sophisticated models over time—three dimensions, with altitude now a factor, and then winds, better resolution, and more—and slowly and steadily chipped away at inaccuracy. AI techniques are augmenting these models or even replacing the physics-based approaches. Simulations are able to push back the forces of chaos theory, allowing predictions further and further into the future.

So, too, with many other realms of experience. Simulation has proceeded to gobble up complexity as models continue to become more sophisticated, with great benefits to each of us. There are models related to nuclear fusion that herald advances in developing reactors that can productively generate and harness atomic forces. There are vast "digital twins" of manufacturing facilities or sprawling technological constructions, such as buildings and airplanes, used to determine the stresses and problems within simulated reality before they become real-world issues. There are even twin earths being developed, models that are supposed to be computational descriptions of the entire planet and that mimic its actual behaviors. The simultaneous behavior of a million stars can be simulated, as well as the extreme long-term future of our own solar system. There are simulations of wildfires. Political violence and turmoil are fodder for simulation, and there have even been models of a thousand years of Vietnam empires. Tsunamis can exist in silico. And don't forget nuclear winter. The outcome of an atomic weapons exchange is not just for the postapocalyptic musings of our fiction writers; we may observe these scenarios in our machines.

These programs perform their simulations via a combination of code and data: code to model the world and data to imbue it with verisimilitude. If we want to understand the weather tomorrow, we need good data on what it is today, and the day before, and the day before that. We also need a reasonable model for how everything should change over time: how air moves in the atmosphere, how it is warmed by the light of the sun, how the terrain affects air movement, and so much more. Due to the power of our computers and the sophistication of the software we can use, the detail of these simulations has ballooned. Some models focus on agents that interact, some use huge collections of mathematical equations cranked through over and over, some extract models from data using artificial intelligence, and many use combinations of these and even other techniques.

Through all of this, are we learning that Laplace is correct? Are we on the verge of understanding every dot and tittle of reality, pouring these facts into our machines and seeing the world as it truly is, pixelated and accurate? We're not there, but I think this is something that many scientists are hoping for.

We see this particularly clearly when it comes to trying to use simulations to understand the nature of history. This kind of work often uses the thinking, seen in novels as diverse as Leo Tolstoy's *War and Peace* and Isaac Asimov's *Foundation*, that argues that history is the accumulation of the massive number of small choices made by each and every one of us, day after day. Every decision matters, but in the same way that every molecule in a gas matters: as part of a large calculation that gives us the total pressure in the volume we're studying. Humans are irrational in consistent ways—and sometimes even rational—and these tendencies allow us to make certain generalizations, aggregating our decisions into a regular curve and shape for how history might flow. As a result, the ability to build synthetic worlds and simulate within them allows us to think about these views of history more rigorously.

For example, there is a domain known as cliodynamics—a mathematical and computational study of history—as well as other work in the social sciences inspired by Asimov's concept of psychohistory, a fictional scientific discipline devoted to understanding the large-scale shape of human behavior and history. The field of cliodynamics employs centuries-long data points along with mathematical models to explore questions around societal collapse, examines the nature of war between empires, and even looks at how large societies formed. Some are mathematical analyses, but some are also simulations. This domain aims to bring the rigors of mathematics and computation to the messiness of the arc of humanity.

Yet, there is a vast quantity of contingency in history, from our own lives—a few seconds' difference in the timing of your own conception could mean you would be an entirely different person—to entire empires laid low by disease. As the novelist Thomas Pynchon wrote, with hints of Lovecraftian madness, "Life's single lesson: that there is more accident to it than a man can ever admit to in a lifetime and stay sane." How can we possibly incorporate all of this contingency and complexity into a digital model?

In all these advances then, we run the risk of hubris. We equate these simulations with the world itself and forget that not only might a perfect correspondence be forever beyond our grasp—let alone true understanding—but prediction is not necessarily even the point of a model.

And that is the key here: while prediction is an obvious reason for simulation, it is not the only one. Yes, throwing some equations into the computer along with some data points, pressing return, and seeing what pops out is compelling. You have captured some of the complexity of the world itself in code and brought it under your control. But there are other reasons we might wish to simulate, from understanding some part of our environment to showing the folly of grappling with complexity. This range of purpose for computational worlds is most clearly seen by looking at simulation games.

SIMULATION'S DELIGHTS AND REASONS

My first experiences of *SimCity* were with the original two-dimensional version, which I played in black and white on my family's Macintosh. Despite the lack of graphical sophistication, the game was addictive as all get-out. I can remember my brother and I running our cities for hours, periodically checking back on them to see if we had accumulated enough cash in our city's coffers from our tax base to build an airport. We tried to make perfect little utopias, ones sculpted to have the correct balance of roads, police stations, and other accoutrements of urban design. We unleashed disasters upon our hapless simulated inhabitants. And we learned about how incentives affected urban development.

The manual that came along with *SimCity* was also a delight, full of information and mini essays about urban design, which I pored over. I even vaguely remember convincing my parents to bring me to the local university library to find books included in the manual's bibliography.

When *SimCity 2000* came out, I was overwhelmed with excitement. This new version of the game was colorful and bright and finally provided a third dimension, rendered at an isometric angle that allowed the entire map to appear 3D.

Decades later, *SimCity 2000* holds up: when I showed it to my daughter, she was impressed by its evocative power. The system underneath the hood is the reason why *SimCity 2000* and its predecessor were so engaging. I wasn't simply building a Potemkin city, placing buildings in specific locations or instructing the inhabitants to manufacture certain products. I was required to use a limited touch. I could construct the city's framework, but my resources were finite. I could create spaces for industry and commerce, but whether they were utilized was a product of complex factors: demand for jobs, proximity to housing. It was a description of the world that was far from inert. I was playing with a toy world that was alive, one that gave me knobs to fiddle with but that required patience and humility in thinking about how it responded to these adjustments.

This was by design. The creator of *SimCity*, Will Wright, developed it as an outgrowth of a number of different ideas. The first, and most basic, was that building a virtual world is fun and satisfying. Wright had discovered that he enjoyed building the environment for a previous game of his, *Raid on Bungeling Bay*, more than the gameplay itself. For there is a delight in creating models. That realization, coupled with delving deeply into the world of system dynamics—the same approach to modeling complex systems that was used in *The Limits to Growth*—gave him the framework for how to create a city-building game.

Many urban planners got into their field by playing *SimCity* when they were kids. The same can hopefully not be said for those who played *Civilization II*. One of the small handful of computer games I became absorbed by, *Civilization II* was another isometrically rendered massive simulation. However, this game involved developing an entire society, creating cities and armies, and slowly conquering the globe as you methodically stepped through the technological and social progress of human history. It was a very different game than *SimCity*, with different kinds of goals. Nevertheless, both games tapped into something fundamental: people love to construct an environment, tweak parameters or choices, and see how they affect an enormously nonlinear world.

These games are far from reality. Cities don't operate according to the rules of Will Wright, and civilizations don't adhere to the ideas of Sid Meier, the creator of *Civilization*. The world is far more complex than that. But playing these games gives one an intuitive set of frameworks for thinking about feedback and systems, and even what better versions of these models could be. As noted in a research paper titled "Six (or So) Things You Can Do with a Bad Model," "If the use of a bad model provides insight, it does so not by revealing truth about the world but by revealing its own assumptions and thereby causing its user to go learn something about the world."

Ken Forbus, a professor of computer science and education at Northwestern University, wrote a short essay in 1996 entitled "Why Computer Modeling Should Become a Popular Hobby." This gem of a paper is full of wonderful turns of phrase and ideas—from "simulation construction kits" to "articulate simulations"—but the upshot is that modeling and simulation are not just necessary skills for understanding the world but also deeply fun activities. There is a joy to building a model that can be examined and tinkered with and that can even bite back (for these are dynamic simulations, of course).

There will always be a gap between reality and simulation, whether hobby or computer game or industrial-grade model. And that's okay. This gap is necessary for avoiding what might be called the kitchen sink conundrum: as more and more detail—both in modeling features and data—is thrown into a simulation, there is no guarantee that it will get us closer to a good understanding of reality itself. This point was articulated decades ago in a RAND Corporation report by David Leinweber. Entitled "Models, Complexity, and Error," the paper examines two types of error: error of specification and compounded error of measurement.

Error of specification refers to how accurate a model is in accounting for the richness of the system being modeled. A more sophisticated model, with more operations on the input, will hopefully correspond better to the real world; it will be more accurate. So, as the complexity of the model is increased, it will adhere better to reality and there will be a lower error of specification. On the other hand, there is also the compounded error of measurement. The more complex a model, the more likely that any measurement error will compound and cause the outputs to become wildly inaccurate.

So one curve goes down with complexity (error of specification), and the other goes up (the errors as a result of measurement). That means, then, that the overall error has a minimum: there is an "optimal model complexity," and, as per the paper, "further complicating the model buys nothing." In fact, recognizing this trade-off goes back even further to a 1968 article, "Predicting Best with Imperfect Data" by William Alonso. The idea of limitations also feels related to certain aspects of chaos theory, and even to how complex models—with oodles of interacting components—yield emergent behavior that is not only hard to simulate but hard to predict.

This trade-off between complexity and accuracy is a humbling realization. Do not add complexity in the hopes of approaching perfection. Not only might there be a diminishing return to this effort, but it could be entirely counterproductive. Computational and mathematical models are powerful, but they must be used carefully. As Leinweber writes, "Weak data require simpler models." Do not succumb to throwing the kitchen sink into a model as the default mode. Understanding itself—something asymptotically approached—is an important goal, as well as prediction. Simulation helps us to understand how different variables interact or how specific parameters affect the output, and the more complex a model, the less likely you are to understand it. Simulation can allow us to see how overall patterns emerge from smaller interactions between the individual parts of a system. Simulation can place a bounding box on our uncertainty, even if it can't predict the world in all its details. Because that might also be a goal forever unreachable. As Ecclesiastes sorrowfully notes, "Indeed, man cannot guess the events that occur under the sun. For man tries strenuously, but fails to guess them; and even if a sage should think to discover them he would not be able to guess them."

Lovecraft creeps back into a world of Laplace, but only if our singular goal is prediction. The computational maps are not the territory; there will always be a bit of Lovecraft in our world.

HUBRIS AND HUMILITY

The Limits to Growth's World3, one of the first computer simulations to penetrate our popular consciousness, was also deeply controversial. Even a half century later, trying to have a discussion about *The Limits to Growth* with a certain group of people—perhaps an intersection of the politically minded and computationally savvy—is no different from trying to have a dispassionate conversation about the best music from your youth. It just isn't possible.

The controversy lies in the alarmist conclusions of the authors and their model. By examining the effects of uncontrolled population growth and economic development—from pollution to resource usage—they contended that humanity would soon bump up against the constraints of the earth, causing catastrophic population decline. But what about advances in technology, you ask? Alas, as per *The Limits to Growth*, no technological panaceas would matter in the long run. If we allow the population and the economy (or, more precisely, "capital") to grow exponentially, no matter the technological advances, there will be some sort of collapse. Another path is therefore necessary. The authors advocated a kind of planetary equilibrium going forward, a stasis in our global population and the economy (this is eliding many details and caveats, and what this actually means and entails is a bit complicated).

Unsurprisingly, then, soon after World3's release, it was swamped with criticism. There were simplifications and inaccuracies, the critics charged. Or the biases of the creators were imposed onto the outcome. Or there were methodological and ideological objections. In fact, an entire book full of nothing but critiques, *Models of Doom: A Critique of the Limits to Growth*, was published just a year later in 1973.

Surprisingly, many of World3's predictions have held up pretty well, despite—or maybe because of—the model's simplicity. For example, our modern world population, given "business as usual," is within a few percentage points of error of its predicted value. In fact, according to a paper published in 2020, the recalibrated version of the model from 2004 fits our current reality quite well. Of course, there are the usual caveats: There were multiple scenarios in the original simulations, so it's too simple to say whether it predicted the "right" future. Recalibration from the early 2000s might also be moving the goal posts. Nevertheless, this elegant model both was accurate in some broad strokes and also provided insight.

This is even more astonishing considering that the creators of World3 were very explicit: they were not trying to make precise predictions but rather "indications of the system's behavioral tendencies." Their model can teach us about the shape of our civilization's behavior, not the details. For example, only a single type of pollutant is considered, and a global and homogeneous population is described, without any semblance of country borders—a massive simplification. But it provides understanding. We can see how feedback plays a role. It can tell us if the model's outputs will oscillate, collapse, or grow rapidly. We can hopefully learn about the nature of humanity's relationship to our planetary home.

With World3, we can see what happens given the assumptions of the model and no changes to our behavior: world population grows until sometime before 2100, when it begins to drop due to nonrenewable resource depletion. But what if we're wrong about the amount of resources? If we take the same model but double our resources, we get similar behavior, albeit a bit time shifted. Are these correct assumptions? Will our behavior change? Is the model correct or is it—as per the criticisms—wildly wrong and nothing more than a prophecy of doom wrapped in some computation? Is this even a fair question to ask of such a stripped-down set of mathematical equations? A simple and incomplete model must wrestle with the fact that our simulation work is never done, but it can still be useful, or at least provocative.

The digital computer has rewarded us with the capacity for creating worlds. Anything we can imagine and describe can be etched into code and simulated. They are Everything Machines. Do we want to use a model of system dynamics, full of stocks and flows? Do we want a set of equations that describes the behavior of an industrial process or nuclear reaction? Do we wish to describe individual agents, interacting according to specific rules and behaviors, and watch complexity emerge? All of this can be encoded into digital machines, allowing us to build any world we want, either far different from our own or quite close.

But the power of creation must be tempered with caution and humility. We must recognize the enormous complexity of reality, which is affected far more by small impacts and outliers than we might wish and which can be particularly difficult to capture in our models. For these simulations are not reality. Even if they are not formally games, they are a kind of make-believe, and we must recognize this. There is an importance to simple models that are understandable, even if imperfect. We can then—slowly and iteratively—make them more complex, but at every stage we must approach this process with a modesty befitting the digital cosmos we are constructing.

In the science fiction author Stanisław Lem's *The Cyberiad*, one of his main characters, a powerful robotic "constructor," sets out to build a toy kingdom for a despot marooned on an asteroid. Out of pity for this exiled ruler without anyone left to rule, the constructor, Trurl, provides him with a realm-in-a-box, a vividly modeled domain complete with simulated subjects over which he may reign. After Trurl demonstrates how to properly turn the knobs to interact with and modify the model world, he leaves the king to it, the ruler's thirst for dominion presumably quenched by this simulacrum.

Later, when Trurl proudly tells his friend Klapaucius of his creation, Klapaucius is horrified: How could you give a world, detailed and accurate, to this despotic monster? He will embitter the lives of these miniature, simulated inhabitants, for lives they truly are. As Trurl built this tiny world to replicate reality, the creatures themselves were real, with all the hopes and fears and suffering possible for all sentient creatures. Happily, when Trurl and Klapaucius go to visit the asteroid, they discover that the king has been overthrown by his simulated subjects, his dead body left as a floating moon to forever orbit the small asteroid.

We are rarely confronted with this situation, where the simulation ends up being more realistic than expected. Too often, we are faced with the opposite, even if we are tempted to forget it: that our simulations of reality are not actually the world around us. They are models—toys for insight—but not to be confused with the real world.

The building of worlds is a gift of computation. It must be used with suitable humility.

———

Now it's time to focus on the connections between computing and one particularly important aspect of the cosmos, at least for humans: life.

10

Bits and Biology

Where Computers and Biology Meet

The world of computing is, alas, rife with malware: computer viruses, Trojan horses, and all manner of computer programs that do nasty things to our machines. In biology, we have something similar: real viruses and harmful bacteria that make us sick or do their best to kill us. But what if we combined these abilities?

First off, the natural question is: Why? What could be gained by taking the worst of one realm and combining it with another? This seems like a recipe for making our lives more worrisome and miserable. But in addition to feeding humanity's insatiable curiosity, it can be valuable to examine how things might go wrong.

Several years ago, a team of researchers at the University of Washington accomplished a mash-up between biology and code. They created a specific genetic sequence that, after being fed into one of the big sequencing machines that are now widely available in labs and biotech start-ups, caused the program processing the sequence data to be "infected" with a destructive piece of software. In other words, when this machine sequenced the text of a specific DNA strand and processed it with certain software, the computer program involved was affected. We know of DNA as malicious code when it comes to viruses and disease, but now we even have DNA as malicious computer code. This nefarious use combined computational ideas with the very nature of biology.

Happily, the blending of these two domains can also be used for good. The ideas that underlie mRNA vaccines, including the vaccines that provide immunity against COVID-19, are based on concepts that sound very much like computer code. These vaccines contain RNA text that is executed by our own bodies, essentially teaching our immune systems what to defend against.

More broadly, the worlds of computing and biology are deeply intertwined, both in concepts and—increasingly—in reality. Through a combination of greater understanding of biology itself and a growth in computing power, these two realms are coming ever closer. In addition, the weirdness of biology is causing us to rethink how weird computing can be, broadening our sense of what computation is. This is that story.

THE MESS AND THE INFORMATION

While the story of biology might be told in many different ways, my focus is on two elements that highlight how biological systems persist, metabolize, and, in general, do their thing: the mess and the information. There is the wet messiness of living things, squishing and squelching their way from generation to generation. Humans are, in the words of an alien creature on *Star Trek*, "ugly giant bags of mostly water." Water is everywhere in biology, providing the medium for biochemical reactions within a cell, playing a role in our circulatory fluids, and acting as a vital ingredient in the goop that makes up our eyeballs.

We leak, we creak, and we make strange noises. We are messes. This is true not only for everything in the animal kingdom but for all of life. Plants live in dirt, crave the sun, and need water. There are fungi that specialize in breaking down dead materials. Life is messy.

But biology is also the story of information. There are sprawling circuit diagrams of the metabolic pathways of the human cell. Our genetic particulars are stored inside the nuclei of our cells in the four-letter code of DNA, which is then copied to another informational substrate known as RNA. This is translated into proteins—folded-up lengths of amino acid molecules—via a genetic code, which explains how every triplet of RNA is decoded into an amino acid. In addition, this genetic code is far from random. There is redundancy and logic and even some optimization made by the fire of evolution. This process is deeply informational. Proteins are the result of converting the two-dimensional string of information contained within DNA into the three-dimensional shape of function. The way that a protein folds dictates its form and as a result what it can do: the shape of its creases inform which other proteins it can connect with and modify, and its structure determines its rigidity or flexibility. In other words, the information of DNA is transcribed and then translated into a string that folds into a three-dimensional shape, its structure dictating its behavior. This is profoundly logical and rife with information.

That being said, most of our DNA does not code just for genes. While there are three billion base pairs in a human genome—the letters of DNA—only about 1 percent is responsible for being turned into proteins. The remainder is responsible for other things, like controlling the creation of these proteins. Truthfully, though, there's a lot that we don't even know what it does yet. But as we look into this information, we can make insightful analogies to computation for many features of biology. For example, the software developer Bert Hubert wrote a delightful description of "DNA seen through the eyes of a coder," wherein he explored biology's source code, how the source is compiled, and much more.

We have both the mess and the information views of biology. How do we reconcile them?

Some of those who engage with biology don't. Different scientists have the luxury of specializing in different parts. Physicians are firmly in the mess camp, for example, contending with the oozing particulars of the human body. At the other extreme, bioinformaticians—scientists who might look at genetic sequences all day—are steeped in computer programs and massive text files packed with biological information.

But often we need to keep both of these facets in our heads because they are deeply intertwined. How did we learn the information view of biology? By a series of messy experiments: spinning biological samples at high speeds using centrifuges, using machines that heat and cool liquids over and over, and carefully placing droplets into tubes. The information is learned from the mess. But even the information view itself is more than a bit of a mess.

What is the interior of a cell? It's basically a vibrating ball pit full of water, small molecules, proteins, RNA, and more, all jostling each other. Due to Brownian motion—the constant oscillations of these objects, dependent on temperature—everything in this ball pit is moving around quickly, bumping into each other over and over. Sometimes these interactions cause something to happen: an enzyme can cleave a protein in two when they touch within the ball pit of the cell.

This wet, warm, jittery jumble—when full of enough molecules—will give way to statistical likelihoods. There are millions of proteins inside the cell of a single bacterium. So, by the laws of probability, certain things are likely to happen, whether it's a string of DNA being transcribed or a molecule passing through a cell membrane. The information pops out.

When it comes to computers, their processing of information also arises from physical structures. But while biology exploits the mess to create information, computers are engineered to avoid this disorder. Integrated circuits are built using clean rooms, and the digital realm of transistors is constructed by thresholding imprecise electrical behavior to provide the ones and zeros of binary. The imperfections are reduced as much as possible. Biology, on the other hand, was not engineered but rather evolved from imperfect parts, so it leans into the physical in operating with the informational.

Nevertheless, there are many fruitful comparisons between the nature of biology and the nature of computers. Biology is deeply modular, just like software, with its components interconnected and reused. The ribosome is a singularly effective piece of open source technology of sorts, a building block used by every cell in existence. Its details vary from organism to organism: think of the different flavors of Unix but for animals or bacteria. Nevertheless, the ribosome is found throughout life, responsible for translating RNA into protein. Just as in computing, open source code is used in application after application.

Similarly, software is able to operate on code itself—a program might modify another piece of code or convert text into a function—and the same kind of thing happens in biology. The code that describes living organisms, DNA, can be converted into proteins but can also be operated on by proteins, with enzymes—themselves derived from DNA—modifying DNA itself. In computer science, this is related to the idea known as homoiconicity: that code can also be treated as data or information, not just as something to be executed.

There is also abstraction in both biology and computing. At the beginning of this book, I examined the edifice of abstraction that all computers are built upon. Biology has the same thing, moving from DNA, to genes and proteins, to entire metabolic pathways, and all the way up to cells and organs in multicellular organisms.

Biology also has circuitry of a kind, with groups of genes that do specific things capable of being turned off and on depending on the situation. Logic gates, the building blocks of computers—how transistors are used to create such operations as AND and OR, which allow for the creation of all features of the modern digital computer—can even be recapitulated in biology. For example, enzymes can operate according to the same kind of logic as transistors, and this allows for the creation of switches and computational machinery, a programming of biology itself. We can design how genes turn off or on. This is the realm of biological engineering, the idea that we can engineer biological parts to do specific tasks.

But even with such biotechnology, we are not yet firmly in the realm of the information; the mess still leaks through.

The programmer Joel Spolsky—the co-creator of Stack Overflow, that online grimoire we encountered earlier—explained the power of abstraction this way: "What is a file system? It's a way to pretend that a hard drive isn't really a bunch of spinning magnetic platters that can store bits at certain locations, but rather a hierarchical system of folders-within-folders containing individual files that in turn consist of one or more strings of bytes." Spolsky wrote this in the context of a discussion on "leaky abstractions," the many times in which abstraction does not work and implementation details rise up and cause us to be aware of the underlying systems, magnetic platters or otherwise. Because often, no matter how hard we try, our abstractions—in biology and computation—will remain leaky.

Proteins might be floppy and wiggly, instead of having a specific rigid shape. DNA might be converted into RNA and then never become protein, and these RNA messages might be used as signals or switches. The circuitry of cells is only similar to that of computers in some of the most superficial senses, with lots of redundancy, complexity, and noise. Many times the information in biology is pretty messy, and the picture I painted above far too simple.

One of the features of living things that must be recognized—and that engineers coming from the world of technology often fail to grasp—is how incredibly complex they truly are. The information cannot yet be understood without thinking about the mess. Do not be led astray by the world of biohacking, for example, where there's this sense that if we just make one simple change to our behavior or to our diet, or if we consume some magical chemical, then we will automatically live longer, think better, sleep less, and get more done. It assumes that living systems are like simple computational ones. But they're not. They might be unbelievably more complicated than anything humanity has engineered. Not to mention the fact that these living systems are the product of millions of years of evolutionary history, which involves balancing lots of different features and trying to optimize them all. Tune anything a bit differently and the entire system might get out of whack.

That's why caffeine can't possibly be only upside: it can help you stay awake and alert, but it also might make you jittery and have to poop. We are tangled jumbles. Not just we humans, but everything in biology is complex and weird and messy. We must have a certain humility in understanding these systems.

Obviously, we have made vast advances in understanding our own bodies, as well as other living things. The four humors of the ancient and medieval world—blood, phlegm, black bile, and yellow bile—have given way to the four nucleotides of DNA. But we are still nowhere close to a complete understanding of biology.

Some of this is related to the shortcomings of the techniques we use to grasp these wet and quivering systems. For example, in 2017, researchers examined what it might be like to use established techniques in neuroscience to try to understand the workings of a simple microprocessor, the kind of chip used in an Atari game system. They came up far short, with suitably humbling consequences.

Even putting aside our techniques, we are grappling with systems that have evolved to work, not ones that have been designed for understanding. As a result, modularity or other standard engineering practices need not always apply to living things, and to assume otherwise is folly. When it comes to biology, it's complexity and exceptions all the way down.

Nevertheless, slowly and fitfully, biologists are making advances in both understanding biology and the ability to engineer living systems according to our wants and desires.

BIOLOGICAL DESIGN

When I was in high school, I took a drafting and design course. The teacher instructed us on how to properly draw a straight line—he exhorted us to rotate the pencil slowly while drawing it along the ruler—as well as how to draft simple blueprints. For our final project, if I remember correctly, we had to design a small cabin.

When designing this all by hand, it forces you to balance requirements and conditions. If I put this bathroom door here, it affects where the toilet can go. And if I put a window on that wall, it prevents the placement of a door. It's a teetering pile of constraints, all of which must be dealt with. On paper, this means lots of erasing, or at least many drafts.

Computers, however, can act as assistants. The right computer-aided design (CAD) software can provide a modifiable medium, turning your design speculations into something that can be easily played with, no erasing or worrying required. CAD software exists for architecture, mechanical engineering, electrical circuit design, and product design. And, you guessed it, for biology.

Biology allows for a whole variety of design problems. For example, want to design a small genetic sequence and then pop it into some bacteria? Well, there are tiny loops of DNA known as plasmids that can be engineered. Plasmid design is a bustling industry, where you can determine the sequence you want and get it manufactured into a plasmid. The company Genentech became a biotech behemoth on the strength of its production of insulin using these techniques: it inserted the genetic sequence for insulin into the bacterium *Escherichia coli* (generally known as *E. coli*), allowing insulin to be generated and collected in large tanks, kind of like the way beer is manufactured.

However, a protein is not just automatically made because it is coded for somewhere in an organism's genome. There is the issue of regulation. So, further engineering might need to be performed to get this right.

The first advances in engineering the regulation of a biological network were achieved a mere twenty-five years ago. The production of a protein that a DNA sequence codes for is determined by a complex network of regulation, involving other proteins that can make the original sequence's transcription and translation into a protein more or less likely. For example, repressor proteins operate by binding to an area before a genetic sequence, known as a promoter region, reducing how much of it is made.

In 2000, the scientists Michael Elowitz and Stanislas Leibler published a paper in *Nature* detailing what they called the repressilator: a combination of three repressor proteins that interact with each other, along with their promoters. Due to the amount of time it takes for these to operate, as well as the amount of time for these proteins to decay within a cell, the scientists were able to engineer a system that oscillated over time, cycling over the course of hours.

We have come a long way from this example. The world of engineered biology is blossoming. There is now computer software available to develop genetic sequences on your desktop, as well as the regulatory features necessary for it to work. You can drag and drop sequences into biological CAD programs, designing and editing bespoke DNA.

But converting digital text into actual DNA for a protein is only one example of design. While it's relatively straightforward to encode a protein into DNA—you "simply" take the string of ACGTs that becomes the amino acids describing the protein and pop it into a cell—if you're trying to build a system that creates other kinds of molecules, matters are far more difficult.

For biology is not just proteins and enzymes. While proteins are vital for life, they are not the only molecules acting within and between cells. There are a whole host of other chemicals found in organisms. These might be small molecules that act as signals between cells, or chemicals that bind to proteins and turn their functionality on or off, or even molecules like caffeine or toxins.

These molecules are in the liquid bath of life but are not part of the traditional path of DNA to RNA to protein. Instead, they are created from other building blocks by proteins. They are constructed via a metabolic pathway: a set of enzymes—proteins that do things—that are responsible for modifying a molecule one after the other, adding or subtracting atoms, combining or removing components.

One way of thinking about this is as a kind of LEGO set, where there are specific operators, each of which can only do one specific thing: adding a brick, combining two sub-modules together, and so on. We need to figure out the right set of operations that will build us the LEGO model that we want.

But this elides the mess. Different operators are not working in sequence. They are all bouncing around in the cell at the same time, working better or worse depending on specific situations and conditions that might be hard to fathom. And these operators might have side effects as well, or even work at cross-purposes.

The task of determining the proteins—the operators—that will yield the final product is a far from trivial problem, and it is less like drawing a pristine diagram and more like debugging a computer program, stepping through it bit by bit and seeing what works and what doesn't.

But it is not just a matter of figuring out the right complement of preexisting proteins and how to combine them correctly to create this metabolic pathway. It also involves creating bespoke proteins, with new behaviors and abilities. So how do we design these novel proteins? You can't just sketch a three-dimensional shape and build it molecule by molecule. Proteins are strings that fold into shapes, and the problem of determining how a string becomes a particular shape is not easy. It entangles text and physics: the buffeting of amino acids by water molecules, the forces between adjacent amino acids. Even amino acids that are far from each other in this string are able to interact through folding and bending. Historically, predicting how a protein folded was accomplished via physics simulations that attempted to model the atoms' interactions. More recently, sophisticated AI techniques are able to infer the ways these proteins fold up and can make predictions. Nevertheless, this is a difficult problem. To give a sense of the complexity of designing biochemical pathways, several years ago, a genetic engineering facility was given the problem of synthesizing ten different molecules, and only six of these were created successfully after three months. This is very much not word processing.

Instead, this is a complex effort of design that involves figuring out how the bits and pieces of proteins are important and can be recombined, incorporating models within artificial intelligence that can better predict the relationship between text and functionality. All of this is in service of building the DNA sequences needed to create whatever target proteins and molecules we want.

Happily, there are software efforts, both commercial and academic, to make every aspect of this process easier, from designing genetic sequences and circuitry, to imagining novel proteins, to even building the complete genetic sequence for an organism. AI allows for better understanding the numerous, complicated interactions across an entire genome, showing how small DNA changes can yield organism-wide effects. We have even begun simulating living cells within computers.

When conducting biomedical experiments, the two main methods of doing so are known as in vivo—in an actual living animal, like a mouse or a monkey—and in vitro—in a test tube or on a petri dish using microorganisms. In vitro is more controlled, but in vivo can be more informative.

But with the advent of the computer, a third option appeared: in silico. You can build a model of biology in a machine and run experiments there. They might be simplified—there is joking about physicists understanding biology by assuming "spherical cows"—but, hopefully, if they capture the relevant aspects of a system, they can provide insight. These kinds of models can vary quite a bit, with mathematical equations for how molecules diffuse within an organism, how a medicine breaks down over time, and even how viruses spread within a population.

Over the past couple years, scientists have been able to build an entire cell—a small one—inside a computer. In 2022, a paper was published that completely modeled a simplified cell, engineered to have a minimal number of genes, accounting for its nutrients, genetic transcription and translation, waste, and everything else, all moving around in a three-dimensional space. This rich molecular model—which involved simulating the behavior of two billion atoms using a vast collection of interlocking equations—is only possible now due to the vast computational power available.

This is all part of the process of turning the mess into the information. It is far from complete, and it must handle the exceptions and complications of biology, but it is steadily happening. An amalgamation of computing and biology is occurring, through a combination of increased biological understanding and computational power.

———

If we take all of this mixing of biology and computation seriously, the final goal, then, is a blending of biology and technology together, a seamless admixture of bits and cells.

We are not there yet. However, we continue to get closer. Tens of trillions of bytes of information—the scale of the digital holdings of the Library of Congress at one point—can be theoretically stored within a drop of DNA. And the process is beginning. Entire books have, in fact, already been encoded into DNA, and there are companies working to use DNA as an information storage method more generally. When a cell with a minimum number of genes was designed and then actually created, the leader of the project is said to have boasted that this was "the first self-replicating species we've had on the planet whose parent is a computer." There are even growing attempts to connect traditional electronics to the messy wetness of biology. Electronics have been incorporated into moths, creating a sort of brain-machine interface for insects, for instance.

A biological computer has been constructed out of bacteria, and the parallel processing nature of biology has been used to demonstrate that biology can itself compute in novel and surprising ways, from slime molds solving optimization problems to DNA being used to factor numbers (a key aspect of code making and breaking).

Better understanding and harnessing of both the mess and the information in biology is opening up new ways of thinking about computation more broadly, part of a burgeoning domain known as unconventional computing, which uses nontraditional mediums. This blending of the mess and the information is teaching us something profound: that computing can actually be much weirder than we might have imagined. Biology might inspire us to broaden our ideas of computation itself.

For instance, perhaps we should expand our idea of technology, with biology providing evidence for "machines as they could be," in the words of Joshua Bongard and Michael Levin, two scientists who are exploring this space. This might involve taking a less reductionist approach to computational systems, incorporating noise into their operations, or scrambling the distinctions between hardware and software.

We might even expand our sense of computing by thinking in terms of "polycomputing," the notion that a computational component can have multiple functions all overlapping together in a single system. This could be a protein that is involved in a huge number of operations based on its changing shape, a brain circuit that has multiple different behaviors, a set of overlapping genes, or even a chemical that is used as both a neurotransmitter and a source of energy. This polycomputing makes it harder to understand and tease apart different functions, but it's a feature all over biology. It upends the more reductionist approach to computation in biology, but it's vital to be aware of if we are interested in broadening our sense of computing's possibilities.

Bongard and Levin—who developed the idea of polycomputing— even argue that biology and computer science are really just two facets of the same domain: information science. The different mediums of biology and computer science both process information, and they should be increasingly blended together.

More and more, when it comes to biology, it's computation— both the information and the mess—all the way down.

———

This is all related to biology as we know it, all the diversity of life that has evolved on our planet over billions of years. What does computing have to say about life as it could be, the "ideas" of life, independent of the messy reality of biology? That's what we'll look at next.

11

Ghosts in the Machine

Artificial Life

A s I write these words, I'm staring at a small, moving, plankton-like creature. It looks like a slightly deformed circle, with a bit of interior complexity, and two taillike appendages on the side, almost like spoilers. It is known as *Orbium unicaudatus*. But I'm not looking into a microscope, observing microbes scooped from the ocean. I'm staring at my computer screen, looking at a computationally generated "creature."

This *Orbium* is one of more than four hundred distinct species that have been discovered in the digital world known as Lenia.

There are organisms that spin around, ones that undulate as they move, ones that pulsate, and others that have vibrating serrated edges. The diversity of Lenia is rich and weird and what you might expect to see if you placed a droplet of pond water onto a slide. But it's all on a machine (and it can easily be run on your web browser). Created by the researcher and engineer Bert Chan in his spare time (Chan has since joined Google), Lenia is less a massive, microcosmic world per se than it is a set of rules that allows all this complexity to unfurl.

It's also an example of the field of artificial life, which studies life through the use of computers and engineering. Or, in the evocative phrase of the artificial life pioneer Chris Langton, of "life as it could be." One of the main tasks of the field of artificial life is to strip life down to its bare essence, see how it ticks, and discover what truly matters. And, in so doing, embody all of this in algorithms.

This sounds at first like a most dispiriting activity—removing all the wonder and complexity from biology and squeezing it into some lines of computer code—but it is anything but that. Artificial life is a vibrant and vital field. Turning life into a project of code means that we can figure out what is truly necessary and what is just fluff. It combines ideas we have previously seen—of biology as code and the crackling power of simulation—allowing us to be bewitched by a provocative hypothesis: that we can build life within machines, or at least simplified simulacra, and thereby learn about life itself. Artificial life is also focused on the engine of life: evolution.

RERUNNING LIFE

The biological diversity that we are faced with when walking through a forest, or combing the depths of the oceans, or exploring jungles, or even spelunking in the fossil record is but a subset of what might actually be possible. Could biology be based on silicon instead of carbon? Or a different sort of body symmetry? Or a new kind of metabolism?

The fact that we have only charted a tiny fraction of the possibilities of life is hinted at by looking at the weirdness of the creatures from hundreds of millions of years ago, from ones that looked like flat plants to sea creatures with teeth at the end of long tubes. Our tree of life, as branching and robust as it might seem, is simply part of a vastly larger tree of possible life.

All around us are clues that if the tape of life on Earth were rerun, it would look very different. The absence of modern dinosaurs—birds aside—is due to a single violent act on the part of the solar system: a massive asteroid impact and its ensuing volcanic eruptions, ashen skies, and resulting No Good, Very Bad Geologic Boundary between the Cretaceous and Paleogene periods. There are founder effects and population bottlenecks too: periods when a small number of individuals of a population survived large-scale death, or were simply the founders of a group of animals in a new place, and evolution had to reckon with a highly skewed population.

This happens even with humans. The founders might have a relatively high rate of a disease. For example, the French Canadian population, when they settled Quebec, brought with them Gaucher's disease. Or a natural disaster can cause the remaining population to have a certain genetic condition, which is said to have happened with the Pacific island of Pingelap in 1775, when, after a typhoon, the survivors were those that had a high rate of a certain kind of color blindness, something that was far from a given.

But despite all of these hints of contingency—that life might have gone in many different directions—evolutionary biology studied over millions of years is very much an observational science. We can imagine all we want about how life might have been different, soaring over the multiverse of biological possibilities, seeing other life-forms and entire ecosystems, with planet after verdant hypothetical planet teeming with life. But we can't rerun history. Artificial life can, however, give us the ability to explore these possibilities. Charting these alternate ecosystems, and the reasons why they might arise, is a feature of this field.

EVOLUTION IS AN ALGORITHM

When you think of evolution, your thoughts come to dinosaurs perhaps, or Lucy and her *Australopithecus* ilk. Maybe you think of missing links or archaeopteryx, trilobites, coelacanths, finches in the Galapagos. Or maybe the ideas of slow and steady change or "Nature, red in tooth and claw." Or perhaps you consider the features of genetics and mutations.

All of this is fair. Generally, we think of evolution and natural selection as something that operates on intricate living things, over plodding and mind-boggling stretches of time. It is that. But it is not a physical force of biology, or something only happening to microbes in the lab or fossils embedded in the exposed geological cliffs of a landscape.

Evolution by natural selection is an algorithm.

Evolution is the logical and inexorable result of certain features of life. It can be described as an algorithm being run by nature itself, but this algorithm can also be run by humans, as we breed everything from cats to crops. It can also be run inside a computer. Evolution by natural selection is essentially selection across a population of entities that have some variation.

What is this selection? Create a population of items, see which ones are the "fittest" according to some criterion, allow those to reproduce more, introduce some variation (mutations and whatnot), and create a new generation. Repeat until you get something impressive.

And that's it. That can embody evolution by natural selection, which means this algorithm is exquisitely suited to computers. Evolution is one of those elegant bits of code we saw earlier, the ones that can unspool huge amounts of diversity and complexity from a small start. And its overall structure—the set of rules behind evolution—was articulated fewer than two hundred years ago. As Darwin noted, "From so simple a beginning endless forms most beautiful and most wonderful have been, and are being evolved."

For example, let's say you are trying to find an equation that fits some data. You have a whole bunch of pairs of data points, x's and y's. When you graph them, it's not a bunch of static, but it seems like some sort of sophisticated squiggle. So what you can do is generate a bunch of random equations that take these x points and convert them into y's.

Initially the random equations will be terrible fits. They will be way off, or far too simple, or, also likely, far too complex and baroque. But here's where the magic of evolution comes in. Once there is the pressure of selection, the equations that are slightly better—but still god-awful fits with our data—get chosen and then modified. And over many, many generations—for remember, *Homo sapiens* was also not evolved in a day—you end up with something good.

There is some magic in this experience of harnessing the engine of biological creation inside a computer. The first time you run these kinds of programs, it feels weird: there is a population of organisms inside your machine, reproducing, recombining, doing all the organic business of life.

The selection mechanism doesn't even need to be automatic. Just as we have selected crops for certain properties over time, humans can be responsible for selection in machines. Based on this idea, the prominent evolutionary biologist Richard Dawkins created a computer program several decades ago focused on what he called biomorphs. Each creature is described by a string of numbers that tells the program how to draw it. These biomorphs don't actually do anything; their entire behavior is their shape and appearance (they often look like bugs). But their evolutionary fitness is determined by the user: the person running the program is shown a sample of biomorphs and is responsible for selecting the one that she likes best. What are your criteria? Whatever you want. You can be a capricious environment. Maybe you're looking for something cute or something that looks like a trilobite. It doesn't matter. You are the cosmic selector. And over time, you can evolve all sorts of creatures.

ARTIFICIAL LIFE

The first attempt to build a system that could study evolution by embodying the features of biology in code occurred almost as soon as the modern digital computer was constructed. As we have seen over and over, this is the way computing worked. Ideas in their embryonic and perhaps underpowered forms were present nearly at the creation of the computer, from insights into biology to simulations to intelligence. This is the slow cooker property of computation: everything was done almost immediately but then took decades to stew. So it was here as well.

The first digital computer with a modern architecture was developed by a team led by John von Neumann at the Institute for Advanced Study (IAS) in the late 1940s and early 1950s. When this machine was operational, another researcher, Nils Barricelli, decided to use it to develop simple machine "organisms." These organisms were strings of numbers that—instead of using a genetic code—used arithmetic to move and change within the memory of the IAS machine, competing to survive from generation to generation in a kind of digital tide pool. Barricelli was interested in running experiments of evolution on a computer, and he developed this one-dimensional world where strings of numbers persisted from one generation to the next. These numerical organisms interacted with each other, with reproduction, mutation, and heredity. Have those features, and you are well on your way to building something evolutionary.

Several decades later, a more sophisticated version of this kind of thing, known as Tierra, was developed by the ecologist Tom Ray. Tierra was a computational petri dish, where the creatures were not just numbers but actually little computer programs. A set of computational instructions was specified, as well as a mechanism for occasional mutation. Biology and computation were fused together far more than the evocative metaphors of the previous chapter: the genetics of each creature in Tierra consisted of strings of instructions for a virtual machine, with each organism competing for processing resources and interacting with the other creatures intimately. Ray programmed a simple replicator, let it loose within the petri dish, and waited.

After a certain amount of time, more efficient replicators developed. But something else arose too: parasites. Snippets of code that used other replicators in order to be copied evolved within the world of Tierra.

All of this sounds very powerful and straightforward. In some ways it is. But our understanding of evolution is still not advanced enough to yield true open-endedness. We can evolve digital organisms that fit certain criteria. We throw a soup of equations into the in silico world of evolution, and out jump ones that adhere to some experimental data (or whatever it is we're looking for). But can we drop some simple computer programs into the engine of natural selection, wait awhile, and fully expect the digital equivalent of lions and tigers and bears to emerge? No, we are adamantly not there yet. For the devil is found deeply in the details. It's very easy for a computational evolutionary system to get stuck, or to spit out something that appears good, but, when you look a bit more deeply, it's overfitted garbage. Over the past few decades, researchers have studied how all of this evolutionary computation operates: What are the best ways to measure the "fitness" of an organism? What kind of encoding of an organism works best? What amount of mutation is ideal?

Other researchers have examined different features of evolution, such as coevolution when two (or more) species evolve together, and as a result there is a sort of escalating arms race, where no species is ever able to "stay still." This is often described by the Red Queen hypothesis, a nod to the Red Queen's comment in Lewis Carroll's *Through the Looking-Glass* that she has to run as fast as she can just to stay still.

In truth, coevolution is an example of a more general process of constantly responding to an ever-shifting environment, in terms of both the physical environment and the surrounding ecosystem. How do we study this? Here, too, scientists can attempt to embody this in code in order to understand this concept.

But there is still so much left to learn. We need to better understand the nature of encoding: How a genotype (the genes) results in a phenotype (the organism's physical form and behavior). Or how mutations allow an organism to move from one phenotype to another. Or how open-endedness itself can arise, or how evolution can emerge without hard coding it into a computer, or how energy constraints affect life.

Convergent evolution—evolution hitting upon the same solutions over and over—is also something that needs to be better understood. Eyes have arisen multiple times in evolutionary history, wings are found in both birds and mammals, and swimming creatures look similar, from dolphins to salmon. One of the hobbies of our biosphere is making crabs: Take a branch of the tree of life, elongate it, and with an inevitability that is an insult to mammals, you'll make a crab evolve. This process, known as carcinization—this making of crabs—has happened at least five times over the course of the earth's history. And these crabs appear from startlingly broad origins across the taxonomy of life.

But it's not fair to evolution to say that it only favors crabs. It also favors trees. When examined carefully, trees have no clear definition and certainly no single taxonomic one. Trees are not a coherent biological category. They are just a solution to a common problem: plants becoming big and tall often end up as trees. Trees and crabs are just some of convergent evolution's answers to the question of how to make life survive on an unforgiving planet.

All of these open questions are grist for the work of artificial life.

———

However, there is so much more in the field of artificial life than just evolutionary questions. For example, another path is that of cellular automata. This category includes Lenia, described briefly at the beginning of the chapter. The most famous of these cellular automata is the mathematician John Horton Conway's Game of Life. Imagine a grid of squares on a computer screen (or even just a really large chess board, because the rules are so rudimentary). Each square can be one of two colors, black or white. If a square is black—or "alive"—it will remain alive as long as two or three of its eight immediate neighbors are also alive. Any more than that, and it becomes white at the next step in time—that is, it dies from overcrowding—and any less than that, it dies from isolation. If a white square is surrounded by exactly three black squares, it is born (and becomes black).

This sounds silly and simple. And also not particularly interesting. But hidden within these rules—or at least hidden within the interactions between the rules and a grid of sufficient size—lies a great deal of complexity. If you run these rules in parallel on a large grid, patterns emerge. Stable structures appear, or ones that blink back and forth. Patterns of squares of a specific shape, called gliders, can even migrate across the grid. The classic glider, for example, looks sort of like a waggling *Tetris* block, shuffling diagonally across the grid. There is a wonderful diversity out of this simple description of the Game of Life. It is yet another example of an emergent microcosm.

Conway's Game of Life first came to the attention of many people through the writer Martin Gardner's Mathematical Games column in *Scientific American* in 1970, and it took readers by storm. Gardner later estimated that millions of dollars' worth of unsanctioned computing time was devoted to simulating the Game of Life after this article was published (this was before the personal computer era, so these resources were at places like universities, companies, or national laboratories).

People working with the Game of Life have come up with new patterns, from a glider gun—a repeating pattern that generates gliders and shoots them off into the distance—to building an entire computer within the world of the Game of Life. Many people have also explored cellular automata more broadly, even modifying their rules and how these systems work. I came to this world late—the 1990s or so—discovering Gardner's writings about the Game of Life through his books that collected his columns. I was entranced by it, downloading software that could implement the Game of Life on our old Macintosh.

Lenia is part of this lineage, swapping out discrete time steps and individual grid points for a smoothed-out time and space and with a suitable modification for how the rules operate (the name Lenia is derived from the Latin word for "smooth"). Bert Chan developed this variation on the Game of Life and then set out to discover what this world contained. Through a combination of watching what would result when random patterns were placed into Lenia, tweaking parameters and shapes, and even a kind of evolutionary selection, Chan eventually discovered hundreds and hundreds of "organisms," patterns that are stable. New versions of Lenia have also been developed that have mechanisms for creating local variation in their rules or that employ the conservation of mass as a means for generating organisms that are not just patterns but a bit more "organismal."

There are still many other directions for the field of artificial life to take. For example, there is a cluster of models known as Particle Life, which embody a kind of simple physics and allow dots to interact according to specific rules. From these simple rules, particles coalesce into organic patterns that move in stunning and lifelike ways. And so much more is being studied.

We have come a long way from ancient and medieval ideas about the nature of life. For some, it was a life force, an animating spirit, a *pneuma* that created a clear dividing line between living and not living. By 1537, one of the big names of alchemy, Paracelsus, described a mechanism for generating a tiny creature known as a homunculus. This process involved combining manure, semen, and blood into a kind of vitalism-inspired smoothie. I assume this did not work.

We are still struggling with the nature of life, from reproduction and metabolism to information persistence and evolvability. We are continuing to explore the complicated blend of information and mess that biology consists of. But artificial life might be able to offer us windows into these ideas.

———

I was trained as a biologist. It's been so long that I'm not sure my membership card is still active, but the world of biology—whether studying the information or the mess—is something for which I have a deep reverence. Biology is unfathomably rich and detailed, from the diversity of species that coat our planet to the shivering clock-work complexity found within a single cell. To build a simulacrum of any of this inside a machine, or even the rules for how it all arises, is something that must constantly mingle humility with hubris.

Nevertheless, the fact that computation is able to touch upon the origins of life, what life might look like on other planets, or even the nature of life itself is worth marveling at. Artificial life is a realm for creation and philosophy. The philosopher Blaise Pascal described humans as thinking reeds: fragile creatures that are nonetheless able to consider the universe. Computers have taken this to a level that would make Pascal weep.

The story of evolution, how it works, how it makes organisms dance from generation to generation, the changes it wreaks—this is the engine that drives the plot of our biosphere. Life, the details it preserves in its information-laden molecules, its chemical reactions, every specific feature it has: all are slowly being encapsulated in code. And with each encapsulation, we learn more about life itself. Whether or not artificial life ever becomes—or is considered—"real" life is one of those questions that I don't think we will answer in the near future. But even so, the very fact that we can experiment with life, learn how its emergent behavior arises, and study all the other properties and definitions—both clear and vague, amorphous and nonnegotiable—entirely within computers is worthy of constant astonishment. We are—a bit, slowly, and haphazardly—as gods.

———

This imperfect and haphazard quality of our godliness is the subject we turn to next.

12

Confronting
the Edges of Software

Bugs, Reality, and Humility

The Talmud is rife with stories and legal disputes. Among many legal scenarios that the Talmud examines—from monetary damages to capital punishment—are those surrounding the ownership of birds when found in the vicinity of a dovecote (their housing structure). For example, if a baby bird is found within fifty cubits of a dovecote—a cubit is a forearm's length or about eighteen inches—then it is the property of the owner of the dovecote. But if it's beyond that, then the finder gets to keep it.

But, asked Rabbi Jeremiah, what happens if one foot of the baby chick is within fifty cubits of the dovecote but the other foot is beyond fifty cubits? The answer is that the other rabbis kicked Jeremiah out of the house of study. He was being annoying and a bit of a smart-ass. When it comes to writing computer programs, however, you have to be like Rabbi Jeremiah, relentlessly thinking about edge cases, weird exceptions to the rule, and rare situations. Otherwise you are going to be beset by glitches and bugs.

Just as science moves forward by engaging with and incorporating the discrepancies between theory and reality, computer programs are improved the same way. Code is not perfect. Far from it. Bugs and glitches are a fundamental feature of computing. In fact, the dirty secret of code is that we will almost never create a perfect piece of software. Debugging is a constant property, teaching us about the world around us, about our own failures to understand it, and more than a little about humility.

Of course, glitches can sometimes be delightful, when they aren't destroying essential systems. When a version of *Microsoft Flight Simulator* was released in the summer of 2020, people were stunned to discover a very narrow and impossibly tall skyscraper in Australia, and that Buckingham Palace was rendered as a massive office complex. These were obviously errors of some sort, but they gave many players a sense of unexpected delight when they explored these places and found obviously incorrect features. The mind-bogglingly complex computer game *Dwarf Fortress* has a changelog for the ages, as its developers root out errors that include "Giraffe is trainable for war" and "Boots don't count as shoes, military gets bad thoughts." There was also a time—before an upgrade was made—when a single carp could become so menacing that it could kill an entire group of dwarves. These errors are so strange—and inadvertently funny—that they point the developer not only toward things to fix but also to the deeply complicated nature of *Dwarf Fortress.*

Ultimately, bugs and glitches are windows into two features of our world: They show us human limitations, as we grapple with the gap between how we think a piece of software works and how it actually does. And they can show how the digital realm is deeply intertwined with the world around us. Computer errors illustrate both how we think about software and how computation is built from—and impinges on—reality.

DEBUGGING AND HUMILITY

Debugging is a strange activity. If you have not written computer code before, you might assume that writing the code is the entirety of the task of creating software. The programmer just sits down, code pours out of her mind, and she runs the finished program. It is not so. Writing code is part of it. Another significant portion of the programmer's time is spent figuring out what libraries to use and how to write specific functions, handle interface weirdness, or deal with any other particularities of the operating system or machine.

But a large fraction of many software developers' time is spent rooting out errors. This is the process of debugging. It can be as simple as discovering that a semicolon has been forgotten at the end of a line or that some curly brace } was misplaced. Many developer environments even help the programmer avoid these missteps. But often, the bug is much more subtle. There is a period of frustration, frustration, and more frustration, followed by enlightenment and insight. And then the developer bumps up against the next bug, and the process repeats.

Debugging has been described as something like the scientific method. And it is, where there is a process of hypothesizing what is wrong—basically having some sort of mental model of why the program is working the way it is—and slowly testing and refining this until the program works correctly.

The difference is that a programmer is not trying to learn something new necessarily. She's simply trying to get the code to work as intended. Sometimes scientific truths are discovered by accident, and that might be closest to what goes on with debugging. But more often it's a long and laborious process that results in learning that the world is exactly as the developer wanted it to be. Nevertheless, the tale of a bug rooted out, when told well, can be as full of twists and turns as the story of slaying a text-based dragon.

In 2002, an email system administrator working at a university shared a story online that sounded impossible: someone couldn't send emails more than about five hundred miles away. To be clear, this is kind of insane, and the person telling the story was well aware of this fact.

But the admin was working with a trained statistician—the chair of a Research Triangle–area university's statistics department—and this statistician had done their homework. The statistics chair worked with a geostatistician to map the failing emails, and they eventually determined that if they sent emails to recipients closer than five hundred miles away, they were delivered. Otherwise, they failed. The statisticians eventually narrowed this distance to about 520 miles.

So what was the problem? After descending into the software running these machines, the administrator realized that a recent change to the servers had downgraded the email system but hadn't changed other parts of the system, causing a time-out failure to happen. And this time-out would be noticed by the remainder of the system after about three milliseconds. Which, based on the speed of light and how far it can travel in three milliseconds, means messages could only go about five hundred miles—a little less than 560 miles to be exact—before failing.

Thus is the process of debugging: engaging with the complex reality of computer code and constantly being humbled by it. This means that, particularly as our technologies get more complex, glitches and failures are an inevitability. In fact, there was a bug in the very first computer program. Lady Ada Lovelace is often considered to be the programmer for the first modern-style digital computer, writing code for a machine designed in the nineteenth century by Charles Babbage, though it was never actually fully built. Lovelace's program from 1843 was designed to calculate certain mathematical quantities known as Bernoulli numbers, but it actually had a small bug in it. Of course, it's easy to understand that she might have made a mistake, since she couldn't run her code and do what is normal for programmers: iteratively and slowly reduce the errors over time. But bugs are a fact of computing life. There is a phrase from the physics literature that feels appropriate for software that seems impervious to all sorts of failures, except, of course, the ones we can't anticipate: "Robust, yet fragile."

To augment the debugging process, then, some software engineers actually inject failure into the system in order to learn. Much as biologists irradiate bacteria to cause mutations, which can then be used for learning more about how these single-celled organisms function, we can introduce our own errors into computer systems—from problematic inputs and weird files to taking certain parts of the system offline—to make them better.

A taste of this kind of injection can be seen in a widespread programming joke that shows the kind of testing an engineer might engage in, and how it can still yield a robust yet fragile system. One version of this joke is as follows: "A software engineer walks into a bar. He orders a beer. Orders 0 beers. Orders 99999999999 beers. Orders a lizard. Orders -1 beers. Orders a ueicbksjdhd. First real customer walks in and asks where the bathroom is. The bar bursts into flames, killing everyone."

You try lots of errors to make sure a program is robust, but it's still only robust to what you've given it. Then something unexpected happens, and the system is horribly destroyed. There are even methods of automatically feeding random inputs into a piece of software to see how it responds (sort of like that engineer at the bar but on steroids). If all this random data doesn't break the program, then we can be more confident that when it engages with the real world of terribly irrational human beings, it won't fail.

But as the inputs to a program become more complex, we are confronted with the sheer impossibility of trying all permutations. The mathematics of combinatorics basically makes the number of conditions that must be tested astronomically large, something we can never hope to fully explore. Randomness can help us to find additional errors—and we can be clever about trying to choose the right kinds of random input—but we will only ever be sampling a tiny subset of potential situations. And that doesn't even include the kinds of failures that don't full-on break our software but just cause it to continue to work in subtly incorrect ways. Those are disturbingly difficult to root out.

There is an epistemological humility here: there are limits to what we can know about the success and failure modes of our software. Bugs and glitches will seep in. They are inevitable, like mosquito bites when camping or dirt on a toddler. No matter what, they will be there. Sometimes they can be like salt and pepper, spicing up an otherwise bland piece of software. Other times, they are half a worm in an apple.

Stories of glitches and bugs are a kind of folklore within the computing world, told both to entertain and delight and with hints of caution. Do not be too certain. Computational pride goeth before a bug, coding certainty before a fall.

We can make inroads into our software, of course, and even use bugs for greater creation. For example, my friend Max Bittker, whom I mentioned back in Chapter 5, refers to some of this process as "domesticating" bugs. Max will take unexpected and anomalous behaviors he discovers in one setting and repurpose them, turning these weird actions into new features elsewhere. (This gives new meaning to the old defensive comment of programmers, "It's not a bug. It's a feature.")

But often, these failures to understand the systems that we build are hints of Lovecraft seeping into our creations, exposing swaths of our ignorance. The computer scientist Fred Brooks wrote in *The Mythical Man-Month* in 1975 that fixing a software problem has a 20 to 50 percent chance of causing yet another bug. We think we understand what we have built but are then proven wrong, over and over.

This is even more true when software intersects with the real world.

COMPUTING AND REALITY

When the messiness of the world is shoehorned into computer programs—whether the weirdness of names or calendars, or different types of characters in languages other than English, or even how we measure the world around us—it massively increases the complexity of software. All these rough edges of reality must be incorporated into a program for it to handle the world in all its variegated strangeness and avoid failure. For example, one major source of bugs is due to what are known as programming falsehoods, beliefs that programmers have about the world that are incorrect, which, when embodied in software, can cause a large number of problems.

Take time, for instance. You might believe that every day has twenty-four hours, but daylight saving time throws this off: one day per year will have an hour more and another will have an hour less. Or you might think that daylight saving time goes from springtime to fall. Not always: in Morocco, if Ramadan occurs in the summer, the country switches back to standard time for that month. Write a web application that involves time, and you're going to have to handle the many quirks and recursively complex edges of how time works. You can't get away with using the simplification of twenty-four hours. Rabbi Jeremiah's obsession with exceptions is necessary for dealing with these falsehoods.

Or maybe you want to write a program that handles people's names. Do not make the assumption that names can't have numbers in them; they can. And some names can be entirely lowercase. Or someone's name might occur in a list of curse words, so you can't automatically filter those out. And so on and so forth.

If you have the last name of Null—a term that essentially means "nothing" in computer languages—you might be unable to interact with computers because the systems view this name as invalid. You won't be able to sign up on websites because you are like a computational ghost, unable to interact with the stuff of code.

Once a program must contend with the fractal complexity of reality, from time zones to prices and currency, programmers must question the world with the pedantic intensity of Rabbi Jeremiah. Otherwise, things will break.

But the conflict between software and the real world, resulting in bugs, can be even more problematic. Occasionally the sheer physicality of computers can leak through, surprising their users.

Just as humans are not merely brains in vats but messy and supremely physical creatures—we think better when we aren't hungry or when we've had a good night's sleep—computers are not abstract Turing machines. They are physical devices, and this fundamental physicality cannot be ignored. For example, the computational signals of email take time to migrate through the wires of the Internet. We think of the Internet as some sort of evanescent gossamer, outside the physical realm. But it's wires and computers and data centers and cooling systems and tubes. It is made of pieces you can touch and visit. So, too, with all computers.

In 2018, a new MRI machine was installed in a hospital and began to make iPhones stop working, along with Apple Watches, but Android phones were just fine. Why? Well, apparently it was due to a minor helium leak in the MRI machine. This leak only affected the specific oscillator used in Apple devices, necessary for chips to keep time and operate. As per one analysis, "Like an incredibly tiny grain of sand, the helium molecules are small enough to get inside the device, physically stop the clock, and turn your phone temporarily into a paperweight." A similar situation is seen when a cosmic ray flips a bit in a computer and causes data to be corrupted or a program to misbehave. Most of us don't want to have to think about the high-speed particles that come from outer space when we are using our word processors, but the digital is born from the physical.

These two realms have been intertwined and causing unexpected interactions as long as computers have been around. A moth got stuck in a Harvard University computer in 1947. In the 1980s, a tape drive stopped working due to a single floor tile.

This tile—made of aluminum and connected to its neighbors by plastic—was slightly warped, and when someone stood on the edge of this tile, it would rub against its neighbor. This rubbing created micro-sparks, which generated electromagnetic interference, in turn causing the tape drive to malfunction.

There is even a known issue with some computer-monitor cables and their use around certain types of office chairs. When people sit on or get up from "gas lift" office chairs, they can cause an electromagnetic spike, which can actually be received by video cables. According to one manufacturer: "If you have users complaining about displays randomly flickering it could actually be connected to people sitting on gas lift chairs." I am flabbergasted.

Google at one point had to track down a problem with internal machines that were overheating, eventually determining that the wheels holding the rack of these servers were crushed, causing the rack to tilt forward and preventing the coolant for these machines from circulating properly. And I read of a delightful effort to determine why someone's Internet Wi-Fi would only work when it rained. The answer: trees had grown to block a Wi-Fi bridge antenna, but rain would weigh the branches down just enough to remove their interference.

There was even a specific Janet Jackson music video, "Rhythm Nation," that would crash certain laptops, because the song had a particular resonance frequency with the hard drive, causing it to fail. This song caused the hard drive to vibrate just right—or really, just wrong—and the machine would stop working.

Even separate from bugs, we are seeing an increase in the computing realm's effects on the real world. Training AI systems uses so much computing power that it can generate as much carbon dioxide as an entire city in a month, and mining cryptocurrency rivals many countries' energy usage. Data centers worldwide consume about 1 percent of all electricity demand, and that figure is only projected to grow. Computing is modifying our very environment, but we sometimes only realize how connected computation is to reality when we see a glitch.

———

Understanding computing is an instance of the naturalist John Muir's point that "when we try to pick out anything by itself, we find it hitched to everything else in the Universe." Glitches teach us about how we measure time, and bugs can interconnect with the laws of physics. They teach us about our software and about the world around us. But they also teach us about ourselves: that we are finite and limited creatures, unable to ever fully ensure the pristine nature of our own technological creations. We aim for computational systems that work perfectly, but their complexity bumps up against the byzantine messiness of humans, as well as our inability to understand all of these features. And so we fail. We create bugs, we fix them, and then we encounter more. There is a humility to be found here: this isn't a defect in our own makeup to be overcome; it is the nature of our humanity.

Imperfection is our fate.

13

Self-Aware Logic

The Simulation Hypothesis

Computation does not consist of disembodied symbol manipulation; it is built upon physics. Just as there are maddening and delightful glitches that arise when physics impinges on our everyday computing—as we saw with leaked helium atoms affecting iPhones in the previous chapter—there can even be cosmic implications for knitting these ideas together.

Twenty-five years ago, the physicist Seth Lloyd tried to tie together computation and fundamental physical limits, such as the speed of light or the gravitational constant. Specifically, Lloyd calculated the maximum amount of computation that could be carried out by a one kilogram, one liter "laptop," in order to learn something about the limits of physics on computation. And what was the answer? Over 10^{50} operations per second. For comparison, as per estimates from the end of 2024, this is well over a nonillion (10^{30}) times faster than the fastest supercomputers available today. That is a very big number. Of course, the side effect is that this machine—or "ultimate laptop," as Lloyd calls it—would be as hot as a thermonuclear explosion. Do not actually put this machine on your lap.

Scientists have gone even further, dwelling on computation in the context of black holes, for example, or what might happen if entire planets were converted to computronium, essentially matter entirely devoted to computation. The nature of computation, when we think grandly enough, is deeply connected to the nature of physics. But what should we do with all this computing power? Calculate pi to precisions that would boggle the mind? Play ever-more sophisticated games? Train superintelligent machines? Or, perhaps, we might build models of the cosmos.

We now get to the true end point of all this talk about reality, physics, and computing smashing together: Just as we can focus on the "realness" of computing—its profound physicality—what if we posit that reality itself is a program, found within a computer, as some kind of grand simulation? What if we could simulate it *all*—I am now gesturing wildly about me, encompassing all that I can, from my office to the universe itself—within a computer? Does this have implications for the intersecting domains of physics and computing, as well as what we think about ourselves?

Physics, reality, and computation fuse in the simulation hypothesis, the idea that reality might be some sort of computer program, and that we are all living within it. This hypothesis, at least on the surface, feels like nothing more than a weird thought experiment, an idea best discussed late at night by college students. But the simulation hypothesis is not just a vessel for storytelling, a kind of factory for pumping out science fiction. It is being seriously explored by philosophers and scientists and is worth examining.

I'd like to focus, then, on what this simulation hypothesis is and what it might say about both the nature of reality and ourselves. For there might be practicalities that unspool from the idea of the universe as simulation. This hypothesis can have implications for how we think about cosmic physics, as well as the meaning of everything and our place within it. In fact, whether or not you believe it to be true, the simulation hypothesis can shed light on the weirdness of the world. Playing with the simulation hypothesis allows you to think about what a glitch in everyday experience might mean and about the true limits of software simulations.

Even more than that, the simulation hypothesis is a cry for mythos. This hypothesis acts as a scientific mechanism for filling a mythology gap in our lives, providing a kind of lore for our technological society, as the other organizing stories that might have once suffused our lives in a more enchanted era no longer have such an abiding allure.

It's now time to explore the implications—and the mythology—of intertwining reality with computation. Let's begin with the nature of the simulation hypothesis.

THE SIMULATION HYPOTHESIS

There is much talk about living in a digital realm and our transition to the metaverse, an online world where we conduct our business, commerce, and play. For some, this is a hoped-for future; for others, it is a kind of horribly dystopian nightmare, a social-media-suffused life injected with anabolic steroids and poison.

But there is a stronger, and different, version of this yet. What if our entire existence is completely computational? What if we are already inside a computer, each of us nothing more than code running within a vast simulation? "Row, Row, Row Your Boat" has come true: life is but a dream, something dreamed by a machine. We are in *SimCity* and just don't realize it.

The simulation hypothesis states that everything in the cosmos, near and far—from stars, planets, and distant galaxies to plants, ocean cyanobacteria, and each of our minds—is all contained within a giant simulation. This is somewhat shocking to consider.

There are a number of complexities with this hypothesis, of course. For example, what of the human mind? For some people, the biggest open question for how a simulation could possibly contain all of reality is that of consciousness. The mind, derived from one of the most complex objects that we are aware of, would need to be contained within a computer. Maybe it would need to be some sort of quantum computer, or have eye-watering amounts of computational power, or have some other condition that we are nowhere near meeting, but it needs to be at least theoretically possible.

To be clear, this is a bit different from intoning that the brain is "just a computer." The brain is distinct from the kinds of machines we have on our desks, and it processes information differently than the way bits move around in a computer. And of course, the brain and the processes of thought have always been compared to the current technologies, from pumps to steam engines. However, with all of that throat clearing, here's the real question: Is it possible to simulate the entire brain inside a computer? My guess is, probably. But if one assumes that it is, many of the complexities boil down to whether the universe could be simulated on a computer of sufficient size, returning us to the kind of estimate at the outset of the chapter.

When we think of the simulation hypothesis, then, we must think in terms of a massively complex simulation. This would require so much computational power it might mean a return to imagined futures from when vacuum tubes ruled the computing earth: computers the size of cities, with people living and working inside these machines, helping to program and maintain them.

Would we need an entire solar system's worth of material to simulate a universe? I have no idea. Would it require a massive coding effort, or are there simple rules that can be run over and over and—through the intricate ways these rules operate and interact—allow for an emergent complexity? Perhaps this could be like some of the artificial life models, some kind of vast cellular automaton but for all of reality. As long as the computing power is available, with the right initial conditions, our *SimCosmos* would mimic the sophisticated phenomena of reality. Or perhaps everything just needs to look complex, but there are lots of simplifications for parts of the universe that we don't normally examine. For example, most extrasolar planets: as long as we don't interact with them, the computer program can just ignore their details.

The argument for the simulation hypothesis in a nutshell is thus: computing power might eventually become sufficiently strong and available for us to be able to simulate the entirety of a universe within it. Assuming that we can get to that point and we are interested in simulating people, then presumably at least one being somewhere in the cosmos—alien or human—has done this. Given the near certainty of it being done, then, and that those within the simulated cosmos could do the same, we end up with the logically ineluctable conclusion that the odds of us being in the "real" universe are infinitesimally small. There are simply too many simulations of reality—given our assumptions—for it to be at all likely that we are in the "base" reality. Quod erat demonstrandum and so on and so forth: we are right now in a simulation.

My own assessment is that it could be true, but in the absence of any clear evidence I don't particularly care whether I'm in a computer program. (Though don't turn me off!) I'm much more interested in how this hypothesis can get us to think about the intersection of reality and computation in weird and provocative ways, rather than dwelling on its specifics and making the truth of this hypothesis of Utmost Importance (that's a whole other issue).

So how can the simulation hypothesis help us think about our world more interestingly? Specifically, if this is all true, what would it mean for us? Would we ever even know? And if so, how?

HACKING REALITY

In 1972, several members of the AI Lab at MIT compiled a memo of the various mathematical insights, hacks, and programming techniques they had been using for the large computers they were working with, from how to calculate certain mathematical functions to how to exchange variables in the programming language Lisp. This was all collected into a memo that is known as HAKMEM, a sort of computational grimoire that we've already briefly encountered.

And buried within this memo is item 154.

Bill Gosper, the author of this item, tries to debunk the notion that "any given programming language is machine independent." Rather, there are differences in how numbers are stored within different computers. And you can discover these differences by doing certain types of mathematics. Specifically, Gosper argued, if you add together more and more powers of two, depending on what happens, you can determine the properties of the specific machine.

The reason that summing powers of two is special—or at least informative—is because of the way binary works. If you add up the powers of two—2^0, 2^1, 2^2, and so forth—you end up with a number that is described in binary as a string of ones. For example:

```
1111111 → 1 + 2 + 4 + 8 + 16 + 32 + 64 = 127
```

Because no machine can handle an arbitrarily large number, if you keep on adding powers of two—basically putting a one in front of your binary number—eventually something will give. How things break can allow you to determine the way the machine stores numbers and how it works. For example, one style is known as "two's complement" and involves a way of storing both positive and negative numbers. If you keep on adding powers of two, eventually the number goes negative.

But note 154 concludes in a surprising direction: "By this strategy, consider the universe, or, more precisely, algebra." Gosper then proceeds to use some mathematical prestidigitation to (inaccurately) imply something about arithmetic involving infinitely large numbers. He then arrives at the result that "algebra is run on a machine (the universe) which is twos-complement."

To be clear, Gosper's mathematics is spurious and presumably done in jest, but even so, it almost feels like Gosper was taking a quick and playful stab at using mathematics to determine something about the very nature of the numerical engine of the cosmos.

Just as Bill Gosper played with how numbers are "stored" in the universe by examining their properties, we might be able to do the same kind of thing for physics more broadly. We can interrogate the features of any simulation we might be inside by looking for specific properties, for the edges of reality.

What would these edges look like? We can see a hint by looking to mathematics.

In modern computers, there's an interesting phenomenon, where if you do certain kinds of seemingly straightforward mathematics, you don't always get the exact answer you might expect. Because of how numbers can be represented, specifically as floating-point numbers, there can be surprising results. For example, if you give a computer the task of calculating $0.9 - 0.8 - 0.1$ you needn't get zero. Instead, you can get something very close, like -0.0000000000000000278. This is one of those situations where our intuition about how numbers should work breaks down in the face of the specific features and idiosyncrasies of the computer itself. These are edge cases for how numbers and mathematics operate within these machines. While a seemingly esoteric feature of computational mathematics, this is something a programmer needs to be aware of when comparing numbers to each other or otherwise using these kinds of equations.

It turns out we can examine the fabric of mathematics itself and find something akin to a floating-point error in the Platonic realm of numbers, what is known as the Borwein integral. Described by a father and his son—both Borweins—it refers to the area under a series of mathematical curves. In this case, these are successive integrals based on a sequence of odd numbers. The first one in the Borwein sequence is exactly equal to pi divided by two. So is the second. Then the third. In fact, the first seven items in the sequence exhibit this pattern, all of which are exactly equal to pi over two. Until the eighth, which is equal instead to pi divided by two, minus a very tiny number (less than one in a billion).

Looking at this, a mathematical result that is no longer exactly equal to a value but instead is a tiny bit off screams floating-point error. Or something similar. Except in this case, there are no approximations or data structures to blame. This is something that crops up in mathematics itself. It turns out that there are specific reasons why the pattern breaks down when it does, and it can be explained by technobabble around Fourier transforms and convolutions. But this is so curious and worthy of suspicion that it was originally viewed as some sort of computer bug.

But it's not. It's just the strangeness of reality.

I don't think that the simulation hypothesis should be reached for when trying to explain Borwein integrals. The world is just weird, and labeling anything we can't quite understand or that violates our intuition as shoddy craftsmanship in the simulation or a glitch in the Matrix doesn't explain anything. It just cuts off understanding (which is why I am gratified to know that there is a reason behind the weirdness of the Borwein integrals, even if I don't fully grasp it).

There are researchers who have tried to find the limits of the simulation that we might find ourselves inside of. Similar to what Gosper was joking about, these researchers have attempted to use the properties of the universe to determine whether we are in a simulation, and, if so, what it might be like.

There's not much here, to be clear, but there are provocations. For example, one research paper used some previous experimental results to argue that, if we are in a simulation, the computer that runs it—whatever substrate it might be and however it works—is not susceptible to some types of "soft errors," the kinds of random bit flips that happen when a particle, such as a cosmic ray, hits the memory. It's not much, and, the author is clear, it could eventually be found to be wrong, but it's a small bit of evidence for something. Other research has tried to understand, using certain aspects of quantum physics, what the structure of the simulation might be, along with its constraints. Though, again, there are often assumptions on top of assumptions for much of this, and it must all be understood within these contexts. I've even seen the argument that, based on quantum physics, a simulation is simply not possible at all (again, dependent on caveats). We really don't have anything much here, though I applaud the efforts of these scientists.

But we don't have to be content with simply searching for these edge cases. We might be able to hack reality based on its computational properties. Imagine a scientist as a kind of computer game player. An avid gamer tries to find the edges of the computerized game world, where things break down. This is both fun and useful for beating a game. It is unlikely that holding down Shift and typing "FUND" onto the keyboard of the universe will give you money like it does in *SimCity*, but maybe there are not only bugs in our universe but Easter eggs. Or a debugging mode, where you can see what's going on in reality, *Matrix*-style.

The novel *Off to Be the Wizard* by Scott Meyer uses this sort of device. The programmer protagonist discovers a massive text file, within which are the parameters and settings of the universe. After messing with his own settings a bit, he travels back to medieval Europe and uses his hacking skills, and knowledge of this text file, to act like a wizard. You know, the usual.

Others have even mused on the idea of "breaking out" of the simulation entirely. In the biblical tale of the Tower of Babel, the city's residents worked together with the aim of reaching the heavens. As the simulation hypothesis has attracted adherents in Silicon Valley, at least two billionaires have tried their own version, working with scientists to break out of reality. Aside from the question of whether this is a profoundly selfish or naive approach to our reality, what would this even look like? Would it be like taking

Neo's red pill and waking up outside the Matrix (unlikely), like shattering the entire simulation and the reality that we find ourselves in (maybe), or like something entirely different? Would we have to start noticing if we occasionally glitched, or if we periodically smashed our heads against large boxes with question marks on them and received a coin or two?

Perhaps breaking out might involve returning to the idea of homoiconicity: equating code and data. One way of breaking software is through something known as code injection, where code is "injected" into another program in an unexpected and unsanctioned way. For example, one mechanism is via a buffer overflow, where a whole bunch of data is input into a computer program, and because it is more than the program expects, that data rewrites other parts of the program as well, since the data is now viewed by the machine as instructions instead of input. Data becomes bug becomes code. A well-known example of code injection involves a situation in *Super Mario World* on Super Nintendo that allows a specific set of motions to be interpreted as instructions, causing the game to have new behaviors. Using this, a player turned *Super Mario World* into the computer game *Flappy Bird*, by, among other things, jumping in certain specific ways.

Maybe this is the way we hack our own world's computer program, as others have speculated. Would this mean that we can reprogram our simulation by jumping around a room in specific locations or by meditating on a specific word? Or uttering specific phrases or performing specific rituals?

In many ways, then, we have come full circle: we started by speculating about computer code as magic, and now, by taking seriously—at least a bit—the idea that we are living in a computational simulation, we might have to think about magic as high-level hacking. I'm not there—and there has been no evidence of anything like this—but show me a powerful incantation and a glitching skyline, and I'm willing to reconsider.

MYTHS AND ESCHATOLOGIES

The simulation hypothesis can be interrogated, as we've seen, via the limits of computational power but also considerations of space-time lattices and brain models. But ultimately, the simulation hypothesis is a myth. For many touched by the idea of the simulation hypothesis, this myth is something that explains the cosmos itself. It is a heady and organizing principle, defined by an aesthetic that might be best described as singularity chic. It's mind-uploading woo combined with AI concerns, sprinkled with the idea that once superintelligent machines are available, Everything Will Change, and we can no longer think about the future because it will all be so different.

The singularity—a combination of superintelligence and human brains running on computers that involves technological growth so rapid that any future prediction is impossible—has been panned as nothing more than the Rapture for nerds, and to a certain degree it is. But it is also an attempt to immanentize the eschaton through technology: to bring a utopian end-time for all humanity into our world.

The simulation hypothesis is one in a series of eschatologies, these visions of the end points of humanity. This one not only dwells on intelligence and the equation of minds and information but connects computing to all of reality. It even requires thinking about who the creators of our simulation might be, why they are running it, and what they want from us (if anything). In that way, it is deeply theological: it is an all-encompassing theory, requires a leap of faith, and has implications for one's salvation. In fact, it even rhymes with more traditional stories that our societies have long told ourselves. As described in *Geek Sublime* by Vikram Chandra, in certain South Asian texts there are concepts of *chiti* and *abhasa*, which involve the idea that there is a unified consciousness that contains everything, with each individual and their features being "simulated" within this consciousness. There appears to be a clear parallel between aspects of these traditions and the simulation hypothesis.

But mythology is an ancient genre, far from scientific explanation. Myths are not textbooks: they are stories to explain humanity's place in the cosmos, or to guide how we think about ourselves, or to provide any other number of lessons, independent of any kind of historical or scientific veracity. The stories of Tantalus and Sisyphus can be entertaining and thought-provoking without me having any concern that running afoul of the gods will send me to a less savory address in the ancient Greek underworld.

The approach I find much more appealing when it comes to the myth of the simulation hypothesis is one of metaphor and figurative thinking. Like those who are obsessed with the simulation hypothesis as a powerful and accurate myth, this less literal approach also has a distinctive aesthetic or vibe. But it involves reveling in the simulation hypothesis as provocative metaphor and idea, rolling it around in one's mind and seeing how it tastes. It involves rejoicing in the physicality of computing and the details that physicists obsess over. It's less about the hypothesis itself and more about what this thought experiment can tell us about our own universe, and about pointing out the rough and worthwhile edges of computing, whether code injection and simulation games or homoiconicity. Viewing the myth of the simulation hypothesis this way involves caring about *SimCity* and not about mind uploading. Or thinking more about the energy requirements of Bitcoin mining and what that might say about the earth's future, and less about meditating on an exquisitely nested set of simulations. It is simulation hypothesis as mechanism for dwelling on the weirdness of the world and its fractal boundaries with computing.

The seventeenth-century philosopher Baruch Spinoza was, depending on whom you ask, either a heretic and an atheist or a "God-intoxicated man." Spinoza's writings analyzed the Bible and defined the divine. Spinoza's God is often considered a pantheistic one, where the universe and God are equated (hence accusations of both divine intoxication and heresy). Intriguingly, the philosopher and novelist Rebecca Newberger Goldstein has described Spinoza's concept of God as one of "self-aware logic." And since Spinoza also identified God with the universe itself, we find ourselves with an intriguing formula: God = Reality = Self-Aware Logic. Continuously operating logic and operations that loop back onto themselves in a self-referential pattern: these are the hallmarks of computing. In some small way, Spinoza—or at least Goldstein's description of his thinking—had hit upon features of the simulation hypothesis.

This might even be related to the idea of computation as the unfolding process of physics itself. There is an idea that the universe is a kind of large, ongoing calculation. When a river flows to the sea, or a marble rolls down a hill, it is exhibiting a behavior that, if we were to mimic it within a computer, would mean executing a complex calculation involving friction, physical forces, and fluid mechanics. When the countless particles in a gas are helter-skeltering about, there is everything from thermodynamics to Boyle's law at work. Or we at least view it that way, if we were to embody these calculations within a computer. But the universe seems able to handle this mathematics all in real time. The "calculations" of the universe appear to be nothing more than the sum total of all the mathematical laws of nature that we have deduced so far, as well as the ones we have yet to learn. And we get all of these calculations essentially for free.

That is the beauty of an analog computer, when done well. Of course, an analog computer requires us humans to determine the right kind of analogue for any specific calculation. If you want fluid dynamics, a wind tunnel could suffice. But if you want to spell-check a document, the universe might not be nearly as obliging in providing the calculation you need. Nevertheless, scientists are making advances in exploiting the natural calculations of our world, using sound-wave interactions, for instance, to mimic neural networks. Whether or not the universe is a simulation, it might very well be mimicked by a massive calculation of sorts, one that we can exploit using sophisticated analog computation.

The cosmos might not be self-aware, but we are the self-aware logic of the universe writ small. Despite being mechanistic objects with properties that can presumably be simulated, we are conscious. As the astronomer Carl Sagan has noted, we have become a mechanism for the universe to study itself, understanding physics, and biology, and even computation.

14

The Wisdom of Computation

The world of computation is an immense one—as I hope I've demonstrated—but it can also sometimes feel distant from the human. The Internet envelops the planet and software can have millions of lines of computer code. Data centers can store enormous quantities of information and consume energy in amounts that are a nontrivial fraction of all the electricity generated by our worldwide civilization.

This book is an attempt to—at least somewhat—reinvigorate the human in the world of computing, from understanding how our machines can help us think better to seeing the similarities between the realm of language and text and the world of code and software. But computing also extends its tendrils into simulating entire societies, the nature of life itself, and the question of whether all of reality is a computer program. And code can make us think we have godlike abilities of wizardry. While I've tried to stress throughout this book that understanding computation should breed a certain humility, this can be difficult.

So let us return to engaging with technologies in ways that are true to our humanity and see what sort of wisdom we can glean. Let's begin with thinking about how technologies change more broadly, something to which we are too often blind.

———

All around me are reminders of obsolete technology. In our home, there are telephone jacks for phones we don't have. There are cables for television we don't use. And in a previous home, there was a central vacuum system that we never even turned on.

Change happens, but we live among the remnants—ruins is probably too strong a term—of these forgotten technologies. Cars have CD players for compact discs few people buy these days. Libraries allow you to borrow DVDs even if you stream all your movies and don't own a DVD player.

I view these less as the inertial burden of technology and more as layered mementos of people's predictions of the future. They are reminders of the inability to recognize that the future will be different from what has come before. Technological change is not a simple extrapolation of what we have now, only a bit faster or sleeker. It will be qualitatively different.

My daughter used to pay very careful attention to my driving. But it wasn't just about road safety or even sightseeing; it was about information. She noticed landmarks and streets, she asked about road signs, and she carefully watched traffic lights. All of this was because, as she happily informed me, she wanted to be ready to drive when she was older. Her excitement for this milestone many years away was infectious, and all this attention and learning was in service of this goal. But all I could think about when she would tell me this was that there was a decent probability that she would never have to learn how to drive. With the developments and buzz around autonomous driving (though this promise continues to recede into the future), there's a chance she may never need to drive herself. Something that requires hours of practice and has been a rite of passage for American teenagers for decades might simply evaporate from our culture. This skill could very well end up going the way of the finer points of mastering the buggy whip. The same could be said—post-pandemic and with the popularity of technology-mediated shopping—for the skill of purchasing items within an actual store.

Or even mastering the dial tone.

In the mid-1950s, to prepare users for the switch from telephone operators connecting calls to rotary dialing, AT&T made a film focused on how to use a rotary telephone: *Now You Can Dial*. The film explains the details of a rotary dial, what a busy signal sounds like, and the nature of the dial tone, among other features. It even, bizarrely, recommends that people "wait for at least a minute, or about ten rings," for someone to answer before hanging up.

But for many younger telephone users, these features are vanishing. If you've only ever used a smartphone, you might never have even heard a dial tone.

The passage of time means inevitable changes in technologies. Some are small: I doubt many people lament the absence of calculator watches or floppy disks. But other changes are far larger. And they don't just provide elements of nostalgia for period pieces on prestige television; they infiltrate numerous aspects of our lives. When one of these technologies evaporates—such as driving a car or owning a telephone that sits on a table or hangs from a wall—it can rewire how we think about the world and our place in it.

These larger technological changes almost swamp our ability to fathom a period beforehand. As someone from the so-called *Oregon Trail* generation, I can remember times before widespread Internet access, how I used the library and its card catalog, how I navigated our telephone books, and how I called a particular phone number to receive the exact time of day when I needed to reset our clocks after a power outage. But I also can't entirely. It's hard to remember—and imagine—what this world was like, because it has been overwritten so thoroughly by the one we currently live in. While I can recall the experience of my local librarian helping me first use a card catalog, I also struggle to remember the specifics of how to navigate one, so completely has my usage of computerized catalogs erased this skill. That world, so close and yet so distant, is difficult to imagine oneself within. It is truly a case where "the past is a foreign country."

We live through technology in time spans, periods of time when something is widespread, taken for granted, and seemingly the natural order of things. Yet these are all ultimately ephemeral.

These technological spans all point to something fundamental about society: even though humanity is really good at dealing with change, we are bad at anticipating it. We are always living within a whole cluster of technological time spans, but too often we are completely oblivious and almost never think about anticipating their ends. Even though we intuitively know that technology is always changing, we still think of individual skills necessitated by categories of technologies as long-lasting or ever-present. We might recognize that cars will improve, but there is a big difference between not having to manually roll down windows or fiddle with a cassette player, and not even having to know how to drive. Information storage methods have changed a lot since magnetic tape and floppy disks, but we are still used to the idea of storing information, even if it's in the cloud. But being able to anticipate when basic skills will be rendered outdated is something we are ill-prepared for. We are in a perpetual state of technological prelapsarianism, forever unable to recognize that the technologies around us—and the society they describe and create—are in a constant state of future obsolescence.

The failure to anticipate the end points of these time spans means that we are certainly living in many of them right now and are simply ignorant. Are we near the end of the skills required to read a map, as Google Maps and its brethren render the knowledge around a map legend no longer necessary? Will the details and etiquette of ringing a doorbell go away, as today's youth simply opt to text when they are outside?

The computer scientist Alan Kay famously noted—provocatively and tongue in cheek—that "technology is anything that wasn't around when you were born." Too often, we're blind to the sheer amount of technology that's all around us if it doesn't have that whiz-bang new feel to it, whether pencils, toasters, books, or even windows. All of this is technology. But so, too, are we blind to the fact that each technology that we're steeped in might be far from permanent.

The world of computation is an extreme version of all of this. Software is inherently transient. Code, in many ways, is fragile and delicate: it can easily fail or become buggy or be rendered obsolete. It's a kind of structured breath, one that can vanish or be deleted at a moment's notice. An online service rarely operates the way it should even just a few years after it is created. Try to emulate an old website and you will be battered by difficulties. Want to see what an old page on the Internet looks like? The Internet Archive might help a bit, but if the web page is interactive, it might not even be possible. I recoil when thinking about how much of what I have written in this book will be obsolete not only within ten years but even by the time the book is printed and finds itself in your hands.

Large software projects evolve over time to address this inherent transience, but I think that wisdom in the realm of computation can only come when we consciously embrace the evanescence of software. As Alan Perlis, the aphoristic computer scientist, noted, "Is it possible that software is not like anything else, that it is meant to be discarded: that the whole point is to see it as a soap bubble?"

With codebases changing over time, and rewriting and updating software occurring as a fact of life, writing code is an exercise in knowing that most of what you will do will be wiped clean by time. Soap bubbles of code will inevitably burst, and that's okay.

Sometimes we see the opposite, when systems that no one ever meant to survive are still around us. For example, Samantha, a friend of mine, has spent time with decades-old COBOL code running on mainframe computers. Her programming task was to convert this old computer code into something more portable and able to be run on modern machines, perhaps because no one ever imagined that it would all last for so long (or, alternatively, maybe they just couldn't imagine it would become obsolete so quickly). This task required delving into the old systems, spelunking into the source code. In some ways, this problem is even worse than the transience: we are stuck with decisions made years ago, in code that might never have been supposed to last. And now we have to contend with the unintended consequences. Whether or not software will last is more a matter of chance and luck, and either way requires absorbing the unknowability of the future of computer code.

In truth, this is the way of all creations of humanity, from cities and bridges to works of art and great books. Time is that slow fire destined to consume everything we have created. Mattering can never be something that happens on some cosmic scale but must be done in the present, for that is all we can be certain will last. The world of computation simply makes this much clearer. All that we create has an expiration date.

———

Alongside this humility, then, we must think about code as something that must be developed in line with our humanity, rather than as something that tries to subvert its nature. The physicist J. Robert Oppenheimer said that "when you see something that is technically sweet, you go ahead and do it and you argue about what to do about it only after you have had your technical success. That is the way it was with the atomic bomb." Technical sweetness can be delightful, but it can also entice us in directions we might not have wished to travel. Sometimes, the power of code can be a dangerous magic, with worrisome ethical and social ramifications to computation.

To reiterate what was explored in the chapter on tools for thought, we must avoid bending ourselves to the needs and whims of our machines. Computation should not narrow our humanity. There are certain features of a rich humanity, whether it's thinking well, spending time with those we love, or enjoying the entertainments of nature and the screen. If our computers are allowing us to avoid these features of life, then we are using technology incorrectly. Technology is not an inherent good, and I don't think everything can be solved by code. The real world is messy and wonderful, and we must be deliberate in how we balance all of this.

Part of this deliberation involves being aware of the limits to the powers of computing. In our era of artificial intelligence, it seems like almost anything is possible. But it is not so.

In another of Stanisław Lem's short stories in *The Cyberiad*, there is a character who wishes to make a machine that is able to compose poetry of singular beauty. But, he soon recognizes, poetry can only derive from a true understanding of an entire civilization: its history, hopes, fears, and culture. So he simulates all of history, recapitulating eons of time, so that some good poetry can be written. Happily, all this work isn't for nothing. The poetry that this machine spits out eventually achieves true greatness.

We have now reached the world of Lem's story: our artificially intelligent large language models swallow up the facts and ideas of our entire civilization, in a quest to generate texts that embody the complexity of our world.

But just as the sun does not truly rise and I'm not actually running rings around you when we are having a debate, language does not always accord with an objective sense of the real world. For example, the semantic embeddings of words and phrases in AI systems tend to incorporate the biases of speech and written text, including stereotypes against women. Professions are viewed as gendered, for example.

But they are also biased in other ways. For at bottom, when we train a model on text, it learns our stories. And narrative is based on what yields a good yarn, one that slices away extraneous or complicating detail. There is always a distinction between the map and the territory. Since these models are ultimately built upon language, we might be forever stuck in the world of stories, simplifying narratives, and tales that lack loose ends or confusing details. Therein lurks Lovecraft. No matter how hard we try, if we use such technologies, we might forever be unable to stare directly at the true nature of the past and our future. Our computers have been jammed full of neat tales, not the bubbling chaos of reality.

Just as we must use technology in line with our humanity and have humility in the face of computing's longevity, we cannot fall prey to an unexamined view of our machines' outputs. Their powers are great but must be interrogated, particularly if we run the risk of outsourcing our storytelling to them.

———

As should be clear by now, computing is a kind of informational attractor: code's particular properties and powers are able to attract—and weigh in on—so many disparate fields, from biology and history to philosophy. As long as it is treated with humility, far from being a pox on our society, computation has the ability to act as a kind of universal solvent for thinking about the world.

Therefore, as we think about how computing fits into our lives, we must make sure that we employ this broad sense of computation and use it to inject wonder and meaning into how we live. Computing can do more than just fill the potholes of life.

In the 1970s, there was an organization known as the People's Computer Company, devoted to educating the general population—including children—about the wonders of computers. It had the following motto: "Computers are mostly used against people instead of for people; used to control people instead of to *free* them; Time to change all that—we need a . . . People's Computer Company."

Since this time, computation has infiltrated every aspect of our society, only a sampling of which is contained in this book. But this charge from the People's Computer Company is still relevant. Is there wonder and delight in how we use our computers? Do they spark omnivorous curiosity? Does computer code evoke the sublime?

Computing is meant to be for humans; all the rest is commentary. Now go and learn.

ACKNOWLEDGMENTS

This is the book that I wish had existed when I was younger and first became really interested in computers. To all of the readers who are in a similar position now: I sincerely hope that this book delights and inspires you, and gets you to think about the human in the world of computing.

Thank you to my agent Howard Yoon, who helped me winnow down the many ideas in my mind into a topic worthy of a book project. Thanks to my editor John Mahaney, as well as the entire team at PublicAffairs, for a herculean effort in turning draft after draft into a book I hope is worth reading. And thanks to everyone at Lux Capital for being so supportive of my book writing; you are a fantastic community and I am lucky to be a part of Lux.

I want to thank my many early readers, who either looked at individual sections of the book or sometimes even subjected themselves to the entire thing. Thanks go to Alice Albrecht, Josh Arbesman, Nadia Asparouhova, Max Bittker, Gordon Brander, Will Byrd, Charles Chamberlain, Ben Fry, Michael Garfield, Chaim Gingold, Mark Glass, Jo Guldi, Samantha John, Ari Kahan, Rohit Krishnan, Steve Krouse, Tony Kulesa, Linus Lee, Corey Maley,

Niko McCarty, Paul Rony, Nahum Shalman, James Somers, Josh Sunshine, Kristoffer Tjalve, Tess van Stekelenburg, Mike Vitevitch, Matt Webb, Olaf Witkowski, and Tarin Ziyaee. I'd also like to thank numerous members of the FLUX Collective for reading chapters and providing invaluable feedback, as well as the members of the Sinai and Synapses working group, whose discussion and feedback were greatly appreciated. In addition, thanks to Hilary McClellen, who did an incredible job fact-checking a book that ranged over far too many topics. All of these people made this book better, but please harbor no ill will toward any of them for the errors that remain; those are mine alone.

In addition, portions of this book appeared in various forms in my newsletter *Cabinet of Wonders*. Thank you to all of my readers for your fabulous feedback and pointers. Bits of various chapters have also been adapted from previously published essays of mine, and such instances are noted in the Chapter Sources and Notes.

Last, I'd like to thank my family: To my wife Debra, for her feedback, encouragement, and especially for her support and immense pride. I am so fortunate to have you as my partner in everything. To my children, for helping me to focus on what is important in the world. And to my parents, for allowing me to see the world of computing from a young age and instilling in me a sense of curiosity and wonder.

CHAPTER SOURCES
AND NOTES

The following is a combination of information on the sources I've used as well as other notes for each chapter. While the sources are not exhaustive, I have worked to include references for all quotations found in the book (full references can be found in the bibliography). Articles and books mentioned by name in the text that are not included in these notes are generally found in the Bibliography as well. For biblical and rabbinic texts, the translations are generally from Sefaria, an incredible online resource. My notes are an attempt to provide clarifications, additional details and context, or some other supplementary information.

INTRODUCTION

Note that I often use computation and computing somewhat interchangeably. Obviously computing can be more than just the computational power of a computer—it can be goofing around with fonts, playing games, word processing, and so forth—but it's also the process of actually computing values and the outputs of algorithms, which is computation. Computation and code as terms might also be interchanged sometimes. Reader, be aware!

Richard Powers described the novel as a "supreme connection machine" in an interview with Jeffrey Williams (1999).

I am not alone in the claim of connecting computing with the humanities and liberal arts. For example, as per Warren Sack in his book *The Software Arts* (2019), the computer scientist Alan Perlis, as early as 1962, felt that computing should be part of a liberal arts education.

The scientist César Hidalgo uses the term "crystallized imagination" in his book *Why Information Grows* (2015). It "rhymes" quite nicely with my idea of software as "spells of crystallized thought."

The point that Robert Alter made in a lecture that I attended is Alter (2023).

Jacob's dawning awareness of the godly nature of his location is in Genesis 28:16.

Portions of this chapter, as well as other chapters, are also adapted from various issues of my newsletter *Cabinet of Wonders*.

1: COMPUTATIONAL WONDER

For details on the early history of digital computing—and its involvement in weather modeling, artificial life, and other simulations—see George Dyson's *Turing's Cathedral* (2012).

The novelist and writer Robin Sloan's essay on artisanal software is Sloan (2020).

I also discuss some of these examples of wonder with computers, including type-in programs, in *Overcomplicated* (Arbesman 2016) and in *The Half-Life of Facts* (Arbesman 2012).

Related to how computers allow us to think about the meaning of humanity, the writer and artist James Bridle noted in *Ways of Being* (2022): "The infinite complexity of computation, which we have divined or dreamed up from the material world, and instantiated in the form of machines, has much to teach us about how we might relate to one another" (page 19).

An exploration of the idea of the "hinge of history" can be found in Fisher (2020).

We might not have technically passed the Turing Test, but for all practical purposes we have achieved the fact that many humans have trouble distinguishing between human-generated text and AI-generated text, AI chatbots can carry on sophisticated conversations, and more. This is an unbelievable achievement.

The point that AI requires space for wonder, rather than just fear, is made in a compelling way by computer scientist Scott Aaronson (2023). Clay Shirky's exploration of the Industrial Revolution and gin consumption is found in Shirky (2010).

2: ALGEBRA AND FIRE

The opening discussion about the Western distaste for the untamed natural world is derived from Johnson (2016), including the descriptions of mountains, particularly pages 260–266.

For Borges and his pairing of algebra and fire, see his short story "Tlön, Uqbar, Orbis Tertius" in Borges (1999) as well as Fishburn (1988).

The discussion of how compilers and interpreters work is necessarily simplified. Compilers can translate between lots of different kinds of languages, and might involve a conversion to assembly code or another high-level language. There are virtual machines—a software kind of processor—byte code, microcode within processors, and other intermediary programs as well. One exploration of all of this complexity and the process of compilers is available in Nystrom (2015–2021).

Further information on the nature and fundamentals of computing can be found in Nisan and Schocken (2021). For the crab computer, see Owano (2012).

The quote from Donald Knuth on computer programs is from Knuth (1982). Edsger Dijkstra's quote is from Dijkstra (1988), though an intriguing counterpoint is found in Guzdial (2020).

The Talmudic story of Moses being shown a rabbi far in his future is from Menachot 29b.

John von Neumann and his not needing anything beyond machine language is found in Auerbach (2018).

The Computer History Museum's source code collections can be found online: "Source Code," Computer History Museum, https://computerhistory.org/playlists /source-code/.

3: DIGITAL ALCHEMY

Leon Kass's discussion of the Tower of Babel is from Kass (2003), pages 242–243. Note that Kass is not happy with this return to a time before Babel.

Much of the discussion of the history and nature of magic is based on Davies (2012).

The Talmud's discussion of creating the world with letters is from Berakhot 55a. The biblical scholar Jacob Milgrom (1990) explores the actions of Moses and God's punishment: Moses used words to try to coerce God, like in ancient pagan magic, something that was unacceptable. In fact, it has been noted that this prohibition was so strong that the work performed in the ancient First Temple in Jerusalem was done in silence.

The nature of the Ilúvatar and Ainur in Tolkien's Middle Earth is from Tolkien (1999). The two quotations from *The Magicians* (Grossman 2010) are on pages 55 and 217, respectively.

For the equation of code and magic, see, for example, Finn (2018). Clive Thompson's discussion of magic and code is from Thompson (2019) on pages 14–16, where he also notes the influences of Dungeons & Dragons and Tolkien. The quotation by Fred Brooks is found in Brooks (1995), pages 7–8. "A Story About 'Magic'" is from Raymond (1998).

The quotation from the Earthsea novels is from Le Guin (1968), page 54.

A discussion of creating a spell-check function in Unix is in Kernighan (2020), along with corrections in the errata online: "*Unix: A History and a Memoir*," last modified May 11, 2022, www.cs.princeton.edu/~bwk/memoir.html.

The book *A Million Random Digits with 100,000 Normal Deviates* is RAND Corporation (1955). *Ajax Penumbra 1969* by Robin Sloan (2014) also discusses the somewhat magical nature of this text. *Numerical Recipes: The Art of Scientific Computing* is Press et al. (2007). HAKMEM is Beeler et al. (1972).

The aspect of sacrifice and magic and artificial intelligence is based on a fascinating insight participants highlighted to me in an Interintellect salon I hosted on "Coding as Magic?" in September 2022. I am indebted to Rob R. for the view of Bitcoin as being sung into existence by computers.

The idea of a devil existing in a print shop during the Renaissance is from Wright (2014).

The discussion of rabbis creating a calf from nothing is found in the Talmud Sanhedrin 67b and is discussed in Trachtenberg (1939), page 84.

While the story of the golem and Gerald Sussman, Marvin Minsky, and Joel Moses has been noted in various sources, I am relying on personal communications with Gerald Sussman (the only one of these three who was still alive during the writing of this book). Also note that Sussman found all of this history in good fun, and did not believe it.

4: OUT OF THE WHIRLWIND

Sources for information about the nature and history of *Spacewar!* are from Levy (2010) and Brand (1974).

tldraw is a good example of a software primitive for infinite canvases, and the creator Steve Ruiz discussed this in the first ten minutes or so of this conversation: Mark McGranaghan and Adam Wiggins, hosts, *Metamuse*, podcast, episode 59, "Infinite Canvases with Steve Ruiz," Muse, July 7, 2022, https://museapp.com/podcast/59 -infinite-canvases/. Note that Lux Capital, where I work, is an investor in tldraw.

The early leaders of the movements for free software and open source software are, respectively, Richard Stallman and Eric S. Raymond. And the public face of one of the most popular open source projects is Linus Torvalds of Linux. At the risk of skipping various stories about these colorful and problematic personalities, I will simply refer you to Eghbal (2020), particularly pages 24 and following, which explore the history behind these people and their movements, as well as how open source culture has shifted over time (among many other topics).

Information on the history and development of Unix can be found in Kernighan (2020). The functionality of Unix has been rewritten multiple times (not necessarily the code itself; it's more like an idea, such as discussed in Stephenson [2003]).

Eghbal (2020) and Stephenson (2003) both discuss the oral and evolutionary tradition of open source computer code. The quotation from *Gods and Mortals* is Johnston (2023), page 4. The quotation from Stephenson (2003) is found on page 88.

The nature of active versus static states of open source (along with the example of the Apollo code) is discussed in Eghbal (2020). See also this quote from Somers (2023): "In 1976, the programmer Richard Stallman created a text-editing program called Emacs that is still wildly popular among software developers today. I use it not just for programming but for writing: because it's open source, I've been able to modify it to help me manage notes for my articles. I adapted code that someone had adapted from someone else, who had adapted it from someone else—a chain of tinkering going all the way back to Stallman."

Research on folktale evolution can be found in da Silva and Tehrani (2016). The book on Cinderella tales is Cox (1893).

The discussion of "Street BASIC" is from Kemeny and Kurtz (1985) on page 56. The quote from Joseph Weizenbaum on dealing with complex systems is from

Weizenbaum (1976). For more on bugs and software complexity, see also Arbesman (2016), where I explore some of these ideas.

God's response when he is overruled by the rabbis is found in the Talmud Bava Metzia 59b. Details on the Zoroastrian fire are from Fisher (2023). The aphorisms of Alan Perlis can be found in Perlis (1982).

Portions of this chapter related to Judaism and long-term thinking first appeared in *Tablet* and are adapted from Arbesman (2021) with permission.

5: A UNIVERSE IN TEN LINES OF CODE

The rabbinic declaration of ten utterances for creation is from Chapters of the Fathers 5:1. Note that there is discussion around these utterances, but one view is it is "And God said" nine times plus the first few words of the Bible itself that are summed up in order to get to ten.

Feynman's quote is from the beginning of Feynman (1996). On the tirelessness of computers, Maeda (2019) discusses some of this further. The details of the game *No Man's Sky* are from Khatchadourian (2015).

Benoit Mandelbrot notes the importance of computers in an interview in the *NOVA* episode "Hunting the Hidden Dimension," first aired on October 28, 2008: "The computer was totally essential; otherwise, it would have taken a very big, long effort." (Transcript available here: www.pbs.org/wgbh/nova/transcripts/3514_fractals .html.)

For the Processing example, note that I am using a relatively recent version of Processing; things might have changed in twenty years.

Related to how Processing and its relatives have re-enchanted programming, this quotation from one of its creators found in Stinson (2021b) is delightful: "What's so fun about stuff like p5.js is that it gets us so much closer to that idea of turning the computer on and Basic is just there. The browser is this sort of lingua franca thing that works everywhere. The idea of having folks be able to connect code right in that environment was super exciting."

For more on the history of Processing, I recommend looking at a two-part oral history in Stinson (2021a; 2021b). The code golf pi calculation can be found here: https://codegolf.stackexchange.com/questions/22009/pi-calculation-code-golf.

For more on Lindenmayer systems and plants, see Prusinkiewicz and Lindenmayer (1990). The quotation about the demoscene and magic of technology is from Viznut (2015), which also cites the concept of "hack value" and its definition.

The book on the snippet of BASIC code is Montfort et al. (2013). For more on the artistic experiments at Bell Labs, see Noll (2016).

Wolfram (2002) also discusses complexity arising from simple computer programs.

6: MACHINE LINGUISTICS

For Robert Alter's thoughts on the language of the Hebrew Bible, see the introduction of Alter (2004). The quote from Vikram Chandra is in Chandra (2014), page 201. Dijkstra's comments on different programming languages are from Dijkstra (1975).

The book *Exercises in Programming Style* by Cristina Videira Lopes (2014), shows, in Python, the numerous styles and genres of programming. Over the course of thirty-three different versions of the same task—to enumerate the most common words in a text ("term frequency")—Lopes demonstrates the many ways of programming within a single language. A more whimsical, and forced, version of this kind of writing is seen in the book *If Hemingway Wrote JavaScript* (Croll 2014). This book tries a little too hard, but I am glad that it exists.

On the history of computing and programming languages, see Petzold (2023). Note that pretty early on, programmers began to write programs in assembly language rather than directly in binary, though each instruction was still generally equivalent to an individual command in binary machine instructions. Lisp as God's own programming language is from Target (2018a), and the Maxwell equations of software idea is from Feldman (2004). The description of Perl is from Wall (1999). Note that "===" in JavaScript is for "strict equality."

For the evolution of programming languages, see Valverde and Solé (2015) and Clancy (2022).

The quotation from the Apple IIe owner's manual is found on page 5 (Apple Computer 1983). One example of a programming language environment that allows the user to loop over dots and create simple animations is tixy.land.

One source for information about Dynamicland—which I never had a chance to visit—is Krouse (2018). Another example of this kind of work is Folk Computer.

I first discovered the quotation by Aristophanes in Nisan and Schocken (2021). For more on the Sapir-Whorf hypothesis and its controversies, see, for example, Pinker (2008), as well as the work of Lera Boroditsky (see her talk at the Long Now Foundation: Lera Boroditsky, "How Language Shapes Thought," recorded October 16, 2010, San Francisco, in Seminars About Long-Term Thinking, Long Now Foundation, https://longnow.org/seminars/02010/oct/26/how-language-shapes-thought/). I also mention my story of kryptonite's color and Vietnamese briefly in Arbesman (2013). The research into the Sapir-Whorf hypothesis for programming is currently in progress by James Evans at the University of Chicago. See also Wexelblat (1981) for some thoughts on Sapir-Whorf and programming languages, as well as how the first coding language one learns might affect one's programming experience.

The computer scientist Edsger Dijkstra (1978) thinks that we need the precision of formalisms—programming languages—even in the face of natural language. So even with AI this might never truly work. Nevertheless, the dream of AI mingled with pseudocode is being worked on, such as one project named SudoLang.

For broader insights into programming languages and code, I recommend Paul Ford's magazine-length essay on code (Ford 2015).

7: IN THE BEGINNING WAS THE SPREADSHEET
This chapter's title is a play on Stephenson (2003).

For more thoughts on creating bespoke computer programs for a small group of users, see Sloan (2020) and Shirky (2004).

For the history of early personal computers, see Nooney (2023). For the history of VisiCalc, a good source is Lohr (2001). Early coverage of VisiCalc can be found in Ramsdell (1980), Green (1980), and an interview with the creators a decade later in Barron (1989). Note that in this last source, Frankston noted that they also referred to it as an "electronic blackboard" (and not as a spreadsheet).

A Small Matter of Programming by Bonnie Nardi (1993) also discusses much around end-user programming and democratization, as well as spreadsheets.

The quote on HyperCard as "an erector set" is found in Goodman (1988) on page xxvi, and the one on bridging the realm of programmers and mouse clickers is from Lewis (1987).

The terms from Seymour Papert about low floors and high ceilings can be found in Resnick (2018) on page 64.

This artificers analogy is from a talk by Alexei Pepers (2019), though the focus in the talk is more on procedural generation, rather than democratization of coding more broadly.

Chaim Gingold (2003) is where the term "magic crayon" is from. The remarks by Steve Jobs on Heathkits is from Berlin (2023). I also discuss both HyperCard and "magic crayons" (and computing gateways) in *Overcomplicated* (Arbesman 2016).

The *Star Trek: The Next Generation* episode that led me to learn about genetics is: *Star Trek: The Next Generation*, season 6, episode 20, "The Chase," written by Joe Menosky, directed by Jonathan Frakes, aired April 26, 1993.

Matt Webb's discussion of how he made his Galactic Compass is in Webb (2024). The "creative collaboration" quotation is from Kottke (2023). The magical invocation by the farmer is from Walton (2020).

One of the foundational computer science textbooks, *Structure and Interpretation of Computer Programs* (Abelson et al. 1996)—known as the Wizard Book due to the appearance of someone who looks like a sorcerer on its cover—in fact notes that the goal of programming is to be understandable by people: "We want to establish the idea that a computer language is not just a way of getting a computer to perform operations but rather that it is a novel formal medium for expressing ideas about methodology. Thus, programs must be written for people to read, and only incidentally for machines to execute."

The Logo advertisement is from *Byte* (1982), page 255.

Portions of this chapter related to HyperCard first appeared in *BBC Future* and are adapted from Arbesman (2019) with permission.

8: TOOLS FOR THOUGHT

See also the book *Tools for Thought* by Rheingold (1985), which is worth looking at.

Others have explored how technology should work with humans, rather than the other way around. For example, from Carse (2012): "To operate a machine one must operate like a machine. Using a machine to do what we cannot do, we find we must do what the machine does. Machines do not, of course, make us into machines when we operate them; we make ourselves into machinery in order to operate them.

Machinery does not steal our spontaneity from us; we set it aside ourselves, we deny our originality."

For more on the design of the keyboard layout, see Wichary (2023).

Thanks to Alice Albrecht for the delightful phrasing of stealing back time from the gods. The phrase "dark illimitable ocean" is in *Paradise Lost*.

Bush's "As We May Think" essay (1945) is about much more than the memex. Nevertheless, it includes a grand vision, which was primarily mechanical and used microphotography and the like, and was based on mimicking the associative connections of the mind.

For more on Paul Otlet, see Wright (2014). And on the brain's work being augmented by tools, see Clark and Chalmers (1998). For a history of word processing, see Kirschenbaum (2016).

The seminal papers of J. C. R. Licklider and Ivan Sutherland are available in Lewis (2021). Engelbart's report is Engelbart (1962).

The quote by John McCarthy on AI is found in Vardi (2012). A good overview of the nature of ChatGPT, as well as word prediction, is Wolfram (2023). A good overview of embeddings, along with some of the examples I use, can be found in Grant Sanderson, "How Large Language Models Work, a Visual Intro to Transformers," YouTube video, 3Blue1Brown, April 1, 2024, www.youtube.com/watch?v=wjZofJX0v4M.

The research on using embeddings in the realm of materials science is in Tshitoyan (2019). The phrase "augmented imagination" is from Brander (2023).

The *Seinfeld* quotation is from: *Seinfeld*, season 4, episode 5, "The Wallet," written by Larry David, directed by Tom Cherones, aired September 23, 1992.

Hofstadter's response to Deep Blue is quoted in Christian (2011) on page 107. Venkatesh Rao's exploration of thought as autotelic is found in Rao (2017). The phrase "epistemic technologies" is seen in Alvarado (2023).

The idea of "bicycles for the mind" (and related phrases) was used by Apple and Steve Jobs over many years, at least as early as 1980, as per Sinofsky (2019).

For "convivial" tools, see Illich (2009). The Berry quotation is from Berry (2000), page 55.

9: LET THERE BE NUMERICAL MODELING

The various creation stories are from Bierlein (1994).

For insights into analog computing (and its form as an analogy), I am indebted to the work of Chaim Gingold (2024) and Corey Maley (2023; 2024).

For *The Limits to Growth* see Meadows et al. (1972).

For "The Call of Cthulhu," see Lovecraft (2014). For "shut up and calculate," see Kaiser (2014). For more on chaos theory, see Gleick (1987, 2008).

For the changes in weather prediction accuracy, see *The Economist* (2023). While I focus on numerical weather prediction, there are AI-based techniques that have the potential to supplant it in certain situations. See, for example, Lam et al. (2023) or Zhang et al. (2023).

For more on the work of Lewis Fry Richardson, see Pontzen (2023). Richardson's quote is in Richardson (1922). Intriguingly the card-reading process and such took most of the time for the ENIAC simulation. It was input-output and manual work that was the bottleneck, as per Edwards (2013), not necessarily Moore's law.

The book *Mirror Worlds* (Gelernter 1992) explores the idea of digital twins and prediction (as well as many fundamental ideas related to computing and more).

For thoughts on history in *War and Peace*, see, for example, pages 821–822 in Tolstoy (2008). A good overview of cliodynamics is found in Turchin (2023). Pynchon's quote is in Pynchon (1963).

Recent insights around "double descent" might mean that adding lots of model complexity can help in some circumstances. A good discussion of the history and design of *SimCity* is in Gingold (2024). The paper "Six (or So) Things You Can Do with a Bad Model" is Hodges (1991). The quote from Ecclesiastes is from Ecclesiastes 8:17.

Models of Doom is Cole et al. (1973). For a discussion of how World3 has fared, see Bardi and Pereira (2022), as well as Herrington (2020). The phrase "indications of the system's behavioral tendencies" is found on page 93 of Meadows et al. (1972).

For more on the complexity of reality and the profound effects of small influences and actions (and our limits to understanding the world due to this), see Klaas (2024).

The Lem story in *The Cyberiad* (2014) is "The Seventh Sally or, How Trurl's Own Perfection Led to No Good." Note that this story was also an inspiration for Will Wright and his creation of *SimCity*, as per Gingold (2024).

10: BITS AND BIOLOGY

The DNA as malicious computer code research is from Ney et al. (2017).

Humans are called "ugly giant bags of mostly water" in *Star Trek: The Next Generation*, season 1, episode 17, "Home Soil," written by Robert Sabaroff, directed by Corey Allen, aired February 22, 1988. Bert Hubert's essay on DNA viewed by a coder is Hubert (2021).

For insight into our modern understanding of biology, I strongly recommend Ball (2023). In addition, McCarty (2023) provides a good overview of the messy and cluttered nature of the cell. Joel Spolsky on leaky abstractions is Spolsky (2002). The research on studying an Atari processor like a neuroscientist is from Jonas and Kording (2017). Related to biology being rife with exceptions, the writer and programmer James Somers has noted of biology, "It's all exceptions to the rule" (2020).

The repressilator is in Elowitz and Leibler (2000). The attempt to synthesize ten molecules using biology is from Casini et al. (2018).

The minimal simulated cell is found in Thornburg et al. (2022).

The quotation on a species "whose parent is a computer" is found in Ball (2023).

One exploration of unconventional computing is Adamatzky and Lestocart (2021). The work by Joshua Bongard and Michael Levin on expanding our notions of machines through biology and on polycomputing is Bongard and Levin (2021; 2023).

Portions of this chapter related to biohacking and biological complexity first appeared in *The Atlantic* and are adapted from Arbesman (2018) with permission.

11: GHOSTS IN THE MACHINE

Information on Lenia can be found online here: https://chakazul.github.io/lenia.html.

Chris Langton's formulation of artificial life can be found in Langton et al. (1992), page xv. Talk of rerunning the "tape of life" seems to have been popularized by the evolutionary biologist Stephen Jay Gould, in his book *Wonderful Life* (1989).

For more on the idea of evolution as an algorithm, see, for example, Dennett (1995). The quotation by Charles Darwin is from the end of *On the Origin of Species*. The biomorphs of Richard Dawkins are discussed in Dawkins (2015), where he also explores evolution as a computer program.

Information on Barricelli's artificial life experiments can be found in Dyson (2012) and Taylor and Dorin (2020). Artificial life is a much broader field than I could describe here, with so many fascinating areas and domains. While there are few recent books on this topic, *Artificial Life* by Steven Levy (1992) is still a good overview, and I used it as a resource for discussing Tierra.

For more on carcinization, see Watson (2023). For convergent evolution of trees, see Ray (2021).

More details on Conway's Game of Life can be found in Gardner (2001).

The details on Paracelsus's recipe for life are from Tripaldi (2022). The phrase "We are as gods" is that of Stewart Brand.

12: CONFRONTING THE EDGES OF SOFTWARE

The Talmudic story about Rabbi Jeremiah is found in Bava Batra 23b.

The five-hundred-mile email story is available online (Harris 2002). The tale of a bug in Lovelace's code is from Target (2018b). The phrase "robust, yet fragile" can be found in Carlson and Doyle (2002). The process of injecting random inputs into software to test it is known as "fuzzing." The fact about fixing software issues causing more problems is from Brooks (1995), originally published in 1975, and is on page 122.

The MRI and iPhone tale is found in Wiens (2018). The tape drive and tile story is found in Thomson (2010). The issue with office chairs and monitor cables is found in a DisplayLink article available online: "Display Intermittently Blanking, Flickering or Losing Video Signal," DisplayLink, https://support.displaylink.com/knowledgebase/articles/738618-display-intermittently-blanking-flickering-or-los. The Google machine overheating story is from McGhee (2020). The Wi-Fi rain story is from Gruevski (2024). And the Janet Jackson music video story is from Chen (2022). Information on data center electricity consumption can be found in Potter (2024).

This chapter has some overlap with the ideas and vibe of my previous book *Overcomplicated* (Arbesman 2016). For a somewhat different take on human limitations, technological complexity, and glitches (along with a bit of discussion of the calendrical weirdness we live with and need to incorporate into our software), I would recommend checking out that book.

Portions of this chapter related to managing technological failures first appeared in *The Atlantic* and are adapted from Arbesman (2024) with permission.

13: SELF-AWARE LOGIC

The "ultimate laptop" calculations are in Lloyd (2000).

The modern version of the simulation hypothesis—the "simulation argument"—can be found in Bostrom (2003). Note that Bostrom's version basically says that a simulation simply needs to be of sufficient complexity for a simulated person. The parts we are not experiencing don't need to be simulated in great detail, including, as Bostrom notes, objects far away in space. Bostrom also discusses "ancestor-simulations" and other ideas that I didn't touch on.

For comparisons of the brain's workings to various technologies, see Marcus (2015).

HAKMEM is Beeler et al. (1972).

A research paper that examines the limits of physics is Alexander (2018), and the research that examines whether we are in a simulation and its relationship to certain aspects of quantum physics is Beane et al. (2014). The argument based on quantum physics against being in a simulation is found in Strogatz (2022). Attempts to break out of the simulation are briefly noted in Friend (2016), and thoughts on hacking our "simulation"—as well as *Super Mario World* and magic—are in Yampolskiy (2023). The computer scientist Scott Aaronson pours cold water on a lot of this, but it's a fun thought experiment, which is sort of where I land (Aaronson 2024).

Another exploration of the simulation hypothesis as myth is found in Deutsch (2021).

Rebecca Newberger Goldstein's comment on Spinoza is found in Goldstein (2006) on page 58. The modern analog computing example and wind tunnel analogy are both found in Wood (2022). Carl Sagan noted in *Cosmos: A Personal Voyage* that "we are a way for the cosmos to know itself."

14: THE WISDOM OF COMPUTATION

The quotation from Alan Kay is found in Greelish (2013). The comparison of software to soap bubbles is from Perlis (1982). Oppenheimer's reference to technical sweetness is from Ratcliffe (2016).

Lem's story in *The Cyberiad* mentioned is "The First Sally (A) or, Trurl's Electronic Bard" (2014). Compare this to the idea that predicting the next token requires a model of the world, as per one of the main scientists who built OpenAI: "what does it mean to predict the next token well enough? . . . it means that you understand the underlying reality that led to the creation of that token." More at BioBootloader (@biobootloader), Twitter (now X), March 28, 2023, https://twitter.com/bio_bootloader/status/1640512444958396416.

You can find information about the People's Computer Company in Markoff (2005) as well as online, such as here: "People's Computer Company & The People's Computers Newsletters," DigiBarn Computer Museum, www.digibarn.com/collections/newsletters/peoples-computer/index.html.

The very last lines of the chapter are an allusion to the words of Hillel in the Talmud Shabbat 31a.

BIBLIOGRAPHY

Aaronson, Scott. 2023. "Should GPT Exist?" *Shtetl-Optimized* (blog), February 22. https://scottaaronson.blog/?p=7042.

Aaronson, Scott. 2024. "Does Fermion Doubling Make the Universe Not a Computer?" *Shtetl-Optimized* (blog), January 29. https://scottaaronson .blog/?p=7705.

Abelson, Harold, Gerald Jay Sussman, and Julie Sussman. 1996. *Structure and Interpretation of Computer Programs*. 2nd ed. Cambridge, MA: MIT Press.

Adamatzky, Andrew, and Louis-José Lestocart, eds. 2021. *Thoughts on Unconventional Computing*. Bristol, UK: Luniver.

Alexander, Samuel. 2018. "A Type of Simulation Which Some Experimental Evidence Suggests We Don't Live In." *The Reasoner* 12 (7): 56.

Alexander, Scott. 2017. *Unsong*. https://unsongbook.com/.

Alonso, William. 1968. "Predicting Best with Imperfect Data." *Journal of the American Institute of Planners* 34 (4).

Alter, Robert. 2004. *The Five Books of Moses: A Translation with Commentary*. New York: W. W. Norton.

Alter, Robert. 2023. "The Challenge of Translating the Bible." Kansas City Public Library, Kansas City, MO, February 16. Video, 1 hr., 23 min. www.youtube .com/watch?v=_vukmd88wpo.

Alvarado, Ramón. 2023. "AI as an Epistemic Technology." *Science and Engineering Ethics* 29 (32).

Andersen, Kurt. 2017. *Fantasyland: How America Went Haywire*. New York: Random House.

Apple Computer. 1983. *Apple IIe Owner's Manual.* Internet Archive. https://archive
.org/details/apple-iie-owners-manual/.

Arbesman, Samuel. 2012. *The Half-Life of Facts: Why Everything We Know Has an
Expiration Date.* New York: Current.

Arbesman, Samuel. 2013. "Some Fun with Color Names." *Wired*, July 3. www
.wired.com/2013/07/some-fun-with-color-names/.

Arbesman, Samuel. 2016. *Overcomplicated: Technology at the Limits of Comprehension.*
New York: Current.

Arbesman, Samuel. 2018. "The Human Body Is Too Complex for Easy Fixes." *The
Atlantic*, April 12. www.theatlantic.com/health/archive/2018/04/biohacking
-siren-song/557849/.

Arbesman, Samuel. 2019. "The Forgotten Software That Inspired Our Modern
World." *BBC Future*, July 23. www.bbc.com/future/article/20190722-the-apple
-software-that-inspired-the-internet.

Arbesman, Samuel. 2021. "The Way-Forward Machine." *Tablet*, November 10.
www.tabletmag.com/sections/community/articles/way-forward-machine-long
-term-thinking.

Arbesman, Samuel. 2024. "What the Microsoft Outage Reveals." *The Atlantic*, July 19.
www.theatlantic.com/ideas/archive/2024/07/microsoft-outage-technological
-systems-fail/679110/.

Asimov, Isaac. 1982. *The Foundation Trilogy.* New York: Del Rey.

Auerbach, David. 2018. *Bitwise: A Life in Code.* New York: Pantheon.

Ball, Philip. 2023. *How Life Works: A User's Guide to the New Biology.* Chicago:
University of Chicago Press.

Bardi, Ugo, and Carlos Alvarez Pereira, eds. 2022. *Limits and Beyond.* London: ExPat
Press.

Barron, Janet. 1989. "Birthing the Visible Calculator." *Byte*, December, 326–328.

Beane, Silas R., Zohreh Davoudi, Martin J. Savage. 2014. "Constraints on the
Universe as a Numerical Simulation." *European Physical Journal A*, 50 (148).

Beeler, M., R. W. Gosper, and R. Schroeppel. 1972. HAKMEM. DSpace@MIT,
MIT Libraries. https://dspace.mit.edu/handle/1721.1/6086.

Berlin, Leslie, ed. 2023. *Make Something Wonderful: Steve Jobs in His Own Words.*
Steve Jobs Archive. https://book.stevejobsarchive.com/.

Berry, Wendell. 2000. *Life Is a Miracle: An Essay Against Modern Superstition.*
Berkeley, CA: Counterpoint.

Bierlein, J. F. 1994. *Parallel Myths*. New York: Ballantine.

Bongard, Joshua, and Michael Levin. 2021. "Living Things Are Not (20th Century) Machines: Updating Mechanism Metaphors in Light of the Modern Science of Machine Behavior." *Frontiers in Ecology and Evolution* 9.

Bongard, Joshua, and Michael Levin. 2023. "There's Plenty of Room Right Here: Biological Systems as Evolved, Overloaded, Multi-Scale Machines." *Biomimetics* 8 (1): 110.

Borges, Jorge Luis. 1999. "Tlön, Uqbar, Orbis Tertius." In *Collected Fictions*. Translated by Andrew Hurley. New York: Penguin.

Bostrom, Nick. 2003. "Are You Living in a Simulation?" *Philosophical Quarterly* 53 (211): 243–255.

Brand, Stewart. 1974. *II Cybernetic Frontiers*. New York: Random House.

Brander, Gordon. 2023. "Everyone Will Have Their Own AI." *Squishy Computer* (blog), March 15. https://newsletter.squishy.computer/p/everyone-will-have -their-own-ai.

Bridle, James. 2022. *Ways of Being: Beyond Human Intelligence*. New York: Farrar, Straus and Giroux.

Brooks, Frederick P., Jr. 1995. *The Mythical Man-Month: Essays on Software Engineering*. 20th anniversary ed. Reading, MA: Addison-Wesley.

Bush, Vannevar. 1945. "As We May Think." *The Atlantic*, July.

Byte. 1982. "Logo: Powerful Ideas in Mind-Sized Bytes." February 1982, 255.

Carlson, J. M., and J. Doyle. 2002. "Complexity and Robustness." *PNAS* 99 (suppl. 1): 2538–2545.

Carse, James P. 2012. *Finite and Infinite Games: A Vision of Life as Play and Possibility*. New York: Free Press.

Casini, Arturo, Fang-Yuan Chang, Raissa Eluere et al. 2018. "A Pressure Test to Make 10 Molecules in 90 Days: External Evaluation of Methods to Engineer Biology." *Journal of the American Chemical Society* 140 (12): 4302–4316.

Chandra, Vikram. 2014. *Geek Sublime: The Beauty of Code, the Code of Beauty*. Minneapolis, MN: Graywolf Press.

Chen, Raymond. 2022. "Janet Jackson Had the Power to Crash Laptop Computers." *Old New Thing* (blog), Microsoft, August 16. https://devblogs.microsoft.com /oldnewthing/20220816-00/?p=106994.

Chiang, Ted. 2002. "Seventy-Two Letters." *Stories of Your Life and Others*. New York: Vintage.

Christian, Brian. 2011. *The Most Human Human: What Artificial Intelligence Teaches Us About Being Alive*. New York: Anchor.

Clancy, Matt. 2022. "Progress in Programming as Evolution." *What's New Under the Sun* (blog), March 10. https://mattsclancy.substack.com/p/progress-in -programming-as-evolution-398#%C2%A7from-simulation-to-reality.

Clark, Andy, and David Chalmers. 1998. "The Extended Mind." *Analysis* 58 (1): 7–19.

Cole, H. S. D., Christopher Freeman, Marie Jahoda, and K. L. R. Pavitt, eds. 1973. *Models of Doom: A Critique of the Limits to Growth*. New York: Universe Books.

Cox, Marian Roalfe. 1893. *Cinderella: Three Hundred and Forty-Five Variants of Cinderella, Catskin, and Cap O'Rushes, Abstracted and Tabulated, with a Discussion of Mediaeval Analogues, and Notes*. London: The Folk-Lore Society. Internet Archive. https://archive.org/details/cinderellathreeh00coxmuoft/.

Croll, Angus. 2014. *If Hemingway Wrote JavaScript*. San Francisco: No Starch.

Da Silva, Sara Graça, and Jamshid J. Tehrani. 2016. "Comparative Phylogenetic Analyses Uncover the Ancient Roots of Indo-European Folktales." *Royal Society Open Science* 3 (1).

Davies, Owen. 2012. *Magic: A Very Short Introduction*. Oxford: Oxford University Press.

Dawkins, Richard. 2015. *The Blind Watchmaker*. Reissue. New York: W. W. Norton.

Dennett, Daniel C. 1995. *Darwin's Dangerous Idea: Evolution and the Meanings of Life*. New York: Simon & Schuster.

Deutsch, L. D. 2021. *Technomythology*. New York: Sacred Bones Records.

Dijkstra, Edsger. 1975. "How Do We Tell Truths That Might Hurt?" EWD 498. EWD Archive, University of Texas at Austin. www.cs.utexas.edu/users/EWD /transcriptions/EWD04xx/EWD498.html.

Dijkstra, Edsger. 1978. "On the Foolishness of 'Natural Language Programming.'" EWD 667. EWD Archive, University of Texas at Austin. www.cs.utexas.edu /users/EWD/transcriptions/EWD06xx/EWD667.html.

Dijkstra, Edsger. 1988. "On the Cruelty of Really Teaching Computing Science," December 2. EWD 1036. EWD Archive, University of Texas at Austin. www .cs.utexas.edu/users/EWD/transcriptions/EWD10xx/EWD1036.html.

Dyson, George. 2012. *Turing's Cathedral: The Origins of the Digital Universe*. New York: Pantheon.

The Economist. 2023. "The High-Tech Race to Improve Weather Forecasting." July 26.

Edwards, Paul N. 2013. *A Vast Machine: Computer Models, Climate Data, and the Politics of Global Warming*. Cambridge, MA: MIT Press.

Eghbal, Nadia. 2020. *Working in Public: The Making and Maintenance of Open Source Software*. San Francisco: Stripe Press.

Elowitz, Michael B., and Stanislas Leibler. 2000. "A Synthetic Oscillatory Network of Transcriptional Regulators." *Nature* 403: 335–338.

Engelbart, D. C. 1962. *Augmenting Human Intellect: A Conceptual Framework*. Menlo Park, CA: Stanford Research Institute.

Feldman, Stuart. 2004. "A Conversation with Alan Kay." *ACM Queue* 2 (9).

Ferguson, Kirby. 2015. "Everything Is a Remix: Remastered." Video, 37 min., 31 sec. www.everythingisaremix.info/everything-is-a-remix-remastered.

Feynman, Richard P. 1996. *Feynman Lectures on Computation*, edited by Anthony J. G. Hey and Robin W. Allen. Reading, MA: Perseus Books.

Finn, Ed. 2018. *What Algorithms Want: Imagination in the Age of Computing*. Cambridge, MA: MIT Press.

Fishburn, Evelyn. 1988. "'Algebra y fuego' in the Fiction of Borges." *Primavera* 12 (3): 383–395.

Fisher, Richard. 2020. "Are We Living at the 'Hinge of History'?" BBC Future, September 23. www.bbc.com/future/article/20200923-the-hinge-of-history -long-termism-and-existential-risk.

Fisher, Richard. 2023. "The Fire That Never Goes Out." Long Now Foundation, April 13. https://longnow.org/ideas/the-fire-that-never-goes-out/.

Forbus, Kenneth D. 1996. "Why Computer Modeling Should Become a Popular Hobby." *D-Lib Magazine*, October. www.dlib.org/dlib/october96/10forbus .html.

Ford, Paul. 2015. "What Is Code?" *Bloomberg Business Week*, June 15–June 28.

Friend, Tad. 2016. "Sam Altman's Manifest Destiny." *New Yorker*, October 3.

Gardner, Martin. 2001. "The Game of Life." In *The Colossal Book of Mathematics*. New York: W. W. Norton.

Gelernter, David. 1992. *Mirror Worlds: Or the Day Software Puts the Universe in a Shoebox . . . How It Will Happen and What It Will Mean*. Oxford: Oxford University Press.

Gingold, Chaim. 2003. "Miniature Gardens & Magic Crayons: Games, Spaces, & Worlds." MS thesis, Georgia Institute of Technology.

Gingold, Chaim. 2024. *Building SimCity: How to Put the World in a Machine*. Cambridge, MA: MIT Press.

Gleick, James. (1987) 2008. *Chaos: Making a New Science*. New York: Penguin.

Goldstein, Rebecca Newberger. 2006. *Betraying Spinoza: The Renegade Jew Who Gave Us Modernity*. New York: Nextbook/Schocken.

Goodman, Danny. 1988. *The Complete HyperCard Handbook*. 2nd ed. New York: Bantam Books.

Gould, Stephen Jay. 1989. *Wonderful Life: The Burgess Shale and the Nature of History*. New York: W. W. Norton.

Greelish, David. 2013. "An Interview with Computing Pioneer Alan Kay." *Time*, April 2. https://techland.time.com/2013/04/02/an-interview-with-computing -pioneer-alan-kay/.

Green, Doug. 1980. "VisiCalc: Reason Enough for Owning a Computer." *Creative Computing*, August, 26–28.

Grossman, Lev. 2010. *The Magicians*. New York: Penguin.

Gruevski, Predrag. 2024. "The Wi-Fi Only Works When It's Raining." *Predrag's Blog*, April 1. https://predr.ag/blog/wifi-only-works-when-its-raining/.

Guzdial, Mark. 2020. "Dijkstra Was Wrong About 'Radical Novelty': Metaphors in CS Education." *BLOG@CACM* (blog), *Communications of the ACM*, November 30. https://cacm.acm.org/blogs/blog-cacm/248985-dijkstra-was-wrong-about -radical-novelty-metaphors-in-cs-education/fulltext.

Harris, Trey. 2002. "The Case of the 500-Mile Email." www.ibiblio.org/harris /500milemail.html.

Herrington, Gaya. 2020. "Update to Limits to Growth: Comparing the World3 Model with Empirical Data." *Journal of Industrial Ecology* 25 (3): 614–626.

Hidalgo, César. 2015. *Why Information Grows: The Evolution of Order, from Atoms to Economies*. New York: Basic Books.

Hodges, James S. 1991. "Six (or So) Things You Can Do with a Bad Model." RAND Corporation.

Hubert, Bert. 2021. "DNA Seen Through the Eyes of a Coder (or, if You Are a Hammer, Everything Looks Like a Nail)." Berthub.eu, January 9. https:// berthub.eu/articles/posts/amazing-dna/.

Illich, Ivan. 2009. *Tools for Conviviality*. Reprint. London: Marion Boyars.

Johnson, Steven. 2016. *Wonderland: How Play Made the Modern World*. New York: Riverhead.

Johnston, Sarah Iles. 2023. *Gods and Mortals: Ancient Greek Myths for Modern Readers*. Princeton, NJ: Princeton University Press.

Jonas, Eric, and Konrad Paul Kording. 2017. "Could a Neuroscientist Understand a Microprocessor?" *PLoS Computational Biology* 13(1): e1005268. https://doi.org/10.1371/journal.pcbi.1005268.

Kaiser, David. 2014. "History: Shut Up and Calculate!" *Nature* 505: 153–155.

Kass, Leon R. 2003. *The Beginning of Wisdom: Reading Genesis.* New York: Free Press.

Kemeny, John G., and Thomas E. Kurtz. 1985. *Back to BASIC: The History, Corruption, and Future of the Language.* Reading, MA: Addison-Wesley.

Kernighan, Brian. 2020. *UNIX: A History and a Memoir.* Published by the author.

Khatchadourian, Raffi. 2015. "World Without End." *New Yorker*, May 11.

Kirschenbaum, Matthew G. 2016. *Track Changes: A Literary History of Word Processing.* Cambridge, MA: Belknap.

Klaas, Brian. 2024. *Fluke: Chance, Chaos, and Why Everything We Do Matters.* New York: Scribner.

Knuth, Donald E. 1982. "The Concept of a Meta-Font." *Visible Language* 16 (1): 3–27.

Kottke, Jason. 2023. "ChatGPT Made Me Cry and Other Adventures in AI Land." Kottke.org, March 29. https://kottke.org/23/03/chatgpt-made-me-cry.

Krouse, Steve. 2018. "The 'Next Big Thing' Is a Room." *Phenomenal World*, October 2. www.phenomenalworld.org/analysis/the-next-big-thing-is-a-room/.

Kuang, Rebecca F. 2022. *Babel.* New York: Harper Voyager.

Lam, Remi, Alvaro Sanchez-Gonzalez, Matthew Willson et al. 2023. "Learning Skillful Medium-Range Global Weather Forecasting." *Science*, November 14.

Langton, Christopher G., Charles Taylor, J. Doyne Farmer, and Steen Rasmussen, eds. 1992. *Artificial Life II.* Redwood City, CA: Addison-Wesley.

Le Guin, Ursula K. 1968. *A Wizard of Earthsea.* Berkeley, CA: Parnassus Press.

Leinweber, David. 1979. "Models, Complexity, and Error." RAND Corporation.

Lem, Stanisław. 2014. *The Cyberiad.* London: Penguin Random House UK.

Levy, Steven. 1992. *Artificial Life: A Report from the Frontier Where Computers Meet Biology.* New York: Vintage.

Levy, Steven. 2010. *Hackers: Heroes of the Computer Revolution.* 25th anniversary ed. Sebastopol, CA: O'Reilly.

Lewis, Harry L., ed. 2021. *Ideas That Created the Future: Classic Papers of Computer Science.* Cambridge, MA: MIT Press.

Lewis, Peter H. 1987. "It's, Well, HyperCard." *New York Times*, August 18.

Lloyd, Seth. 2000. "Ultimate Physical Limits to Computation." *Nature* 406: 1047–1054.

Lohr, Steve. 2001. *Go To: The Story of the Math Majors, Bridge Players, Engineers, Chess Wizards, Maverick Scientists, and Iconoclasts—the Programmers Who Created the Software Revolution*. New York: Basic Books.

Lopes, Cristina Videira. 2014. *Exercises in Programming Style*. London: Chapman and Hall/CRC.

Lovecraft, H. P. 2014. *The New Annotated H. P. Lovecraft*. New York: Liveright.

Maeda, John. 2019. *How to Speak Machine*. New York: Portfolio.

Maley, Corey J. 2023. "Analogue Computation and Representation." *British Journal for the Philosophy of Science* 74 (3): 739–769.

Maley, Corey J. 2024. "Computation for Cognitive Science: Analog Versus Digital." *WIREs Cognitive Science* 15 (4): e1679.

Marcus, Gary. 2015. "Face It, Your Brain Is a Computer." *New York Times*, June 27. www.nytimes.com/2015/06/28/opinion/sunday/face-it-your-brain-is-a-computer.html.

Markoff, John. 2005. *What the Dormouse Said: How the Sixties Counterculture Shaped the Personal Computer Industry*. New York: Penguin.

McCarty, Niko. 2023. "Biology Is a Burrito." *Asimov Press*, February 28. www.asimov.press/p/burrito-biology.

McGhee, Steve. 2020. "Finding a Problem at the Bottom of the Google Stack." *Google Cloud Blog*, March 13. https://cloud.google.com/blog/products/management-tools/sre-keeps-digging-to-prevent-problems.

Meadows, Donella H., Dennis L. Meadows, Jørgen Randers, and William W. Behrens III. 1972. *The Limits to Growth*. New York: Universe Books.

Meyer, Scott. 2014. *Off to Be the Wizard*. Seattle: 47North.

Milgrom, Jacob. 1990. "Magic, Monotheism, and the Sin of Moses." In *The JPS Torah Commentary: Numbers*. Melrose Park, PA: Jewish Publication Society.

Montfort, Nick, Patsy Baudoin, John Bell et al. 2013. *10 PRINT CHR$(205.5 +RND(1)); : GOTO 10*. Cambridge, MA: MIT Press.

Nardi, Bonnie A. 1993. *A Small Matter of Programming: Perspectives on End User Programming*. Cambridge, MA: MIT Press.

Nelson, Ted. 1974. *Computer Lib/Dream Machines*. Published by the author.

Ney, Peter, Karl Koscher, Lee Organick, Luis Ceze, and Tadayoshi Kohno. 2017. "Computer Security, Privacy, and DNA Sequencing: Compromising Computers with Synthesized DNA, Privacy Leaks, and More." Presented at the USENIX Security Symposium, Vancouver, BC. https://dnasec.cs.washington.edu/dna -sequencing-security/.

Nisan, Noam, and Shimon Schocken. 2021. *The Elements of Computing Systems.* 2nd ed. Cambridge, MA: MIT Press.

Noll, Michael. 2016. "Early Digital Computer Art at Bell Telephone Laboratories, Incorporated," *Leonardo* 49 (1): 55–65.

Nooney, Laine. 2023. *The Apple II Age: How the Computer Became Personal.* Chicago: University of Chicago Press.

Nystrom, Robert. 2015–2021. "A Map of the Territory." In *Crafting Interpreters.* Published by the author. https://craftinginterpreters.com/a-map-of-the-territory .html.

Owano, Nancy. 2012. "Scientists Work Up a Crab-Powered Computer." Phys.org, April 14. https://phys.org/news/2012-04-scientists-crab-powered.html.

Pepers, Alexei. 2019. "A Guide to Proc Gen Practitioners." Roguelike Celebration, San Francisco, October 5. Video, 31 min., 24 sec. www.youtube.com /watch?v=mYMdMAvTHpo.

Perlis, Alan J. 1982. "Epigrams on Programming." *ACM SIGPLAN Notices* 17 (9): 7–13.

Petzold, Charles. 2023. *Code: The Hidden Language of Computer Hardware and Software.* Redmond, WA: Microsoft Press.

Pinker, Steven. 2008. *The Stuff of Thought.* New York: Penguin.

Pontzen, Andrew. 2023. *The Universe in a Box: Simulations and the Quest to Code the Cosmos.* New York: Riverhead.

Potter, Brian. 2024. "How to Build an AI Data Center." *Construction Physics* (blog), June 10. www.construction-physics.com/p/how-to-build-an-ai-data-center.

Press, William H., Saul A. Teukolsky, William T. Vetterling, and Brian P. Flannery. 2007. *Numerical Recipes: The Art of Scientific Computing.* 3rd ed. Cambridge: Cambridge University Press.

Prusinkiewicz, Przemyslaw, and Aristid Lindenmayer. 1990. *The Algorithmic Beauty of Plants.* New York: Springer-Verlag.

Pynchon, Thomas. 1963. *V.* Philadelphia: J. P. Lippincott & Co.

Ramsdell, Robert. 1980. "The Power of VisiCalc." *Byte,* November, 190–192.

RAND Corporation. 1955. *A Million Random Digits with 100,000 Normal Deviates.* New York: Free Press.

Rao, Venkatesh. 2017. "Intelligence Reconsidered." *Contraptions* (blog), October 6. https://contraptions.venkateshrao.com/p/intelligence-reconsidered.

Ratcliffe, Susan, ed. 2016. *Oxford Essential Quotations*. 4th ed. Oxford: Oxford University Press.

Ray, Georgia. 2021. "There's No Such Thing as a Tree (Phylogenetically)." *Eukaryote Writes Blog*, May 2. https://eukaryotewritesblog.com/2021/05/02/theres-no-such-thing-as-a-tree/.

Raymond, Eric S. 1998. *The New Hacker's Dictionary*. 3rd ed. Cambridge, MA: MIT Press.

Resnick, Mitchel. 2018. *Lifelong Kindergarten*. Cambridge, MA: MIT Press.

Rheingold, Howard. 1985. *Tools for Thought*. Cambridge, MA: MIT Press.

Richardson, Lewis Fry. 1922. *Weather Prediction by Numerical Process*. London: Cambridge University Press. Internet Archive. https://archive.org/details/weatherpredictio00richrich/weatherpredictio00richrich/.

Sack, Warren. 2019. *The Software Arts*. Cambridge, MA: MIT Press.

Shirky, Clay. 2004. "Situated Software." *Clay Shirky's Writings About the Internet* (blog), March 30. Archived February 11, 2012, at https://web.archive.org/web/20120211075149/http://shirky.com/writings/situated_software.html.

Shirky, Clay. 2010. *Cognitive Surplus: Creativity and Generosity in a Connected Age*. New York: Penguin Press.

Sinofsky, Steven. 2019. "Bicycle for the Mind." *Learn by Shipping* (blog), August 11. https://medium.learningbyshipping.com/bicycle-121262546097.

Sloan, Robin. 2014. *Ajax Penumbra 1969*. London: Atlantic Books.

Sloan, Robin. 2020. "An App Can Be a Home-Cooked Meal." Robinsloan.com, February. www.robinsloan.com/notes/home-cooked-app/.

Somers, James. 2020. "I Should Have Loved Biology." Jsomers.net. https://jsomers.net/i-should-have-loved-biology/.

Somers, James. 2023. "Whispers of A.I.'s Modular Future." *New Yorker*, February 1. www.newyorker.com/tech/annals-of-technology/whispers-of-ais-modular-future.

Spolsky, Joel. 2002. "The Law of Leaky Abstractions." *Joel on Software* (blog), November 11. www.joelonsoftware.com/2002/11/11/the-law-of-leaky-abstractions/.

Stanley, Kenneth O., and Joel Lehman. 2015. *Why Greatness Cannot Be Planned: The Myth of the Objective*. Cham, Switzerland: Springer.

Stephenson, Neal. 2003. *In the Beginning . . . Was the Command Line*. New York: Perennial.

Stinson, Liz. 2021a. "Processing: The Software That Shaped Creative Coding." *AIGA Eye on Design*, October 28. https://eyeondesign.aiga.org/processing-the-software-that-shaped-creative-coding/.

Stinson, Liz. 2021b. "Processing: The Software That Shaped Creative Coding, Part II." *AIGA Eye on Design*, November 1. https://eyeondesign.aiga.org/an-oral -history-of-processing-part-two/.

Strogatz, Steven. 2022. "What Is Quantum Field Theory and Why Is It Incomplete?" *Quanta*, August 10. Transcript. www.quantamagazine.org/what-is-quantum -field-theory-and-why-is-it-incomplete-20220810/.

Target, Sinclair. 2018a. "How Lisp Became God's Own Programming Language." *Two-Bit History* (blog), October 14. https://twobithistory.org/2018/10/14/lisp .html.

Target, Sinclair. 2018b. "What Did Ada Lovelace's Program Actually Do?" *Two-Bit History* (blog), August 18. https://twobithistory.org/2018/08/18/ada-lovelace -note-g.html.

Taylor, Charles. 2018. *A Secular Age*. Cambridge, MA: Belknap.

Taylor, Tim, and Alan Dorin. 2020. *Rise of the Self-Replicators: Early Visions of Machines, AI and Robots That Can Reproduce and Evolve*. Cham, Switzerland: Springer.

Thompson, Clive. 2019. *Coders: The Making of a New Tribe and the Remaking of the World*. New York: Penguin.

Thomson, Patrick. 2010. "The Best Debugging Story I've Ever Heard." *Djinn and Juice* (blog), December 28. https://patrickthomson.tumblr.com/post/2499755681 /the-best-debugging-story-ive-ever-heard.

Thornburg, Zane R., David M. Bianchi, Troy A. Brier et al. 2022. "Fundamental Behaviors Emerge from Simulations of a Living Minimal Cell." *Cell* 185 (2): 345–360.

Tolkien, J. R. R. 1999. *The Silmarillion*. 2nd ed. London: HarperCollins.

Tolstoy, Leo. 2008. *War and Peace*. Translated by Richard Pevear and Larissa Volokhonsky. New York: Vintage Classics.

Trachtenberg, Joshua. 1939. *Jewish Magic and Superstition: A Study in Folk Religion*. New York: Behrman's Jewish Book House.

Tripaldi, Laura. 2022. "Artificial Lives: On the Occult Origins of Chemistry and the Stuff of Life." *MIT Press Reader*, May 18. https://thereader.mitpress.mit.edu /artificial-lives-on-the-occult-origins-of-chemistry/.

Tshitoyan, Vahe, John Dagdelen, Leigh Weston et al. 2019. "Unsupervised Word Embeddings Capture Latent Knowledge from Materials Science Literature." *Nature* 571: 95–98.

Turchin, Peter. 2023. *End Times: Elites, Counter-Elites, and the Path of Political Disintegration*. New York: Penguin.

Valverde, Sergi, and Ricard V. Solé. 2015. "Punctuated Equilibrium in the Large-Scale Evolution of Programming Languages." *Journal of the Royal Society Interface* 12 (107). http://doi.org/10.1098/rsif.2015.0249.

Vardi, Moshe Y. 2012. "Artificial Intelligence: Past and Future." *Communications of the ACM* 55 (1).

Viznut. 2015. "Bringing Magic Back to Technology." *Countercomplex* (blog), April 9. https://countercomplex.blogspot.com/2015/04/bringing-magic-back-to-technology.html.

Wall, Larry. 1999. "Perl, the First Postmodern Computer Language." LinuxWorld Conference and Expo, San Jose, CA. Transcript. www.wall.org/~larry/pm.html.

Walton, Kathryn. 2020. "Farming with Charms in the Middle Ages." Medievalists.net. www.medievalists.net/2020/06/farming-with-charms-in-the-middle-ages/.

Watson, Clare. 2023. "Evolution Keeps Making Crabs, and Nobody Knows Why." *Science Alert*, June 15. www.sciencealert.com/evolution-keeps-making-crabs-and-nobody-knows-why.

Webb, Matt. 2024. "New App! A Compass That Points to the Centre of the Galaxy." *Interconnected* (blog), February 15. https://interconnected.org/home/2024/02/15/galactic-compass.

Weizenbaum, Joseph. 1976. *Computer Power and Human Reason: From Judgment to Calculation*. San Francisco: W. H. Freeman.

Wexelblat, Richard L. 1981. "The Consequences of One's First Programming Language." *Software—Practice and Experience* 11: 733–740.

Wichary, Marcin. 2023. *Shift Happens*. Published by the author.

Wiens, Kyle. 2018. "iPhones Are Allergic to Helium." *iFixit*, October 30. www.ifixit.com/News/11986/iphones-are-allergic-to-helium.

Williams, Jeffrey. 1999. "The Last Generalist: An Interview with Richard Powers." *Cultural Logic*. https://ojs.library.ubc.ca/index.php/clogic/article/download/192194/188958/220043.

Wolfram, Stephen. 2002. *A New Kind of Science*. Champaign, IL: Wolfram Media.

Wolfram, Stephen. 2023. "What Is ChatGPT Doing . . . and Why Does It Work?" *Stephen Wolfram Writings* (blog), February 14. writings.stephenwolfram.com/2023/02/what-is-chatgpt-doing-and-why-does-it-work.

Wood, Charlie. 2022. "How to Make the Universe Think for Us." *Quanta*, May 31. www.quantamagazine.org/how-to-make-the-universe-think-for-us-20220531/.

Wright, Alex. 2014. *Cataloging the World: Paul Otlet and the Birth of the Information Age*. Oxford: Oxford University Press.

Yampolskiy, Roman. 2023. "Hacking the Simulation: From the Red Pill to the Red Team." Preprint, ResearchGate, January 12. https://philarchive.org/rec /YAMHTS-2.

Zhang, Yuchen, Mingsheng Long, Kaiyuan Chen et al. 2023. "Skilful Nowcasting of Extreme Precipitation with NowcastNet." *Nature* 619: 526–532.

INDEX

Credit: Akiva Fleischmann

Samuel Arbesman is Scientist in Residence at Lux Capital. In addition, he is an xLab senior fellow at Case Western Reserve University's Weatherhead School of Management and a research fellow at the Long Now Foundation. He is also the author of *Overcomplicated: Technology at the Limits of Comprehension* and *The Half-Life of Facts: Why Everything We Know Has an Expiration Date*, and his writing has appeared in such places as the *New York Times, Wall Street Journal, The Atlantic,* and *Wired,* where he was previously a contributing writer. He lives in Cleveland with his family. The first computer he used was a Commodore VIC-20.